Tesi di perfezionamento in Neurobiologia sostenuta l'8 giugno 2000

Francesco Mattia Rossi
Laboratoire de Neurobiologie Moléculaire et Cellulaire
Ecole Normale Supérieure
46 rue d'Ulm
75230 Paris Cedex 05, France
e-mail: rossi@wotan.ens.fr

A Study on Nerve Growth Factor (NGF) Receptor Expression
in the Rat Visual Cortex: Possible Sites and Mechanisms of NGF Action
in Cortical Plasticity

Francesco Mattia Rossi

A Study on Nerve Growth Factor (NGF)
Receptor Expression in the Rat Visual Cortex:
Possible Sites and Mechanisms
of NGF Action in Cortical Plasticity

TESI DI PERFEZIONAMENTO

SCUOLA NORMALE SUPERIORE
2000

ISBN: 978-88-7642-280-5

ACKNOWLEDGEMENT

I want to thank all my friends and colleagues at the Scuola Normale Superiore and at the Institute of Neurophysiology CNR of Pisa who helped me during my PhD training period both at a cultural and technical level.

I also thank Prof. M. Raiteri, G.B. Bonanno and R. Sala of the Department of Experimental Medicine of Genova whose scientific contribution to this work has been fundamental.

TABLE OF CONTENTS

INTRODUCTION

DISCUSSION

MATERIALS AND METHODS

INTRODUCTION

INTRODUCTION

Summary

Neuronal plasticity is the term generally used to describe a great variety of changes in neuronal structure and functions, in particular activity-dependent, prolonged functional changes, accompanied by corresponding biochemical and possibly morphological alterations. The first insights about neuronal plasticity have been obtained on simple forms of life, such as *Aplysia californica*. Today, many areas in the mammalian Central Nervous System (CNS) are under investigation, in particular the cortex and the hippocampus. In these areas, neuronal plasticity is considered to be at the basis of learning and memory.

The study of brain plasticity in sensory systems has been particularly fruitful, because these systems can be experimentally manipulated in a very precise manner by changes in the sensory input. The visual system in particular received much attention in the past, mainly thanks to Nobel prizes D.H. Hubel and T.N. Wiesel (1998). These scientists characterised the anatomy and the physiology of the mammalian visual system and paved the avenue for the study of brain plasticity, as described below.

Anatomical and physiological organisation of the mammalian visual system

Three main structures are involved in visual perception in mammals: the eyes, a relay thalamic nucleus (dorsal lateral

1

geniculate nucleus – dLGN) and the primary visual cortex (occipital cortex, Oc1). Figure 1 shows a simple draft of the mammalian visual system structure. Axons of retinal ganglion cells leave the eye grouped in a bundle of fibres, the optic nerve, and reach the dLGN. A percentage of optic nerve fibres from each eye (depending on the animal species, 95% in the rat) cross controlaterally at the level of the optic chiasm, so that each dLGN is functionally connected with both eyes. The inputs from the two eyes remain segregated in the dLGN, in that fibres from the two eyes terminate in alternating eye specific layers that are strictly monocular. This eye specific anatomical distribution is maintained also at cortical level; indeed, geniculate axon terminals project mainly to layer IV of the Oc1, where they are still segregated in alternating eye specific patches (ocular dominance columns). From here on, neuronal projections become more complex and are no more segregated.

The final and correct anatomical organisation of this structure is the result of a fine process that takes place shortly before birth and during early postnatal life in mammals. At birth, all these projections are already present, but, at geniculate and cortical level, axon terminals are still intermixed with each other. The segregation of fibres in eye specific domains takes place first in the dLGN, and successively in the visual cortex.

Cortical cells have specific functional properties that correspond to the peculiar anatomical organisation of neuronal projections. D.H. Hubel and T.N. Wiesel were the first to show that

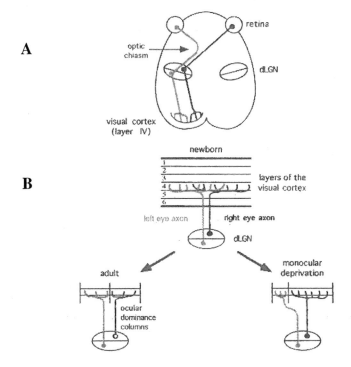

A

B

Figure 1. Schematic diagram of the developing mammalian visual system. (A) The main structures in the visual system pathway are the eyes, the dorsal lateral geniculate nucleus (dLGN) and the primary visual cortex. Axons from retinal ganglion cells form the optic nerve fibres, which cross contralaterally at the level of the optic chiasm (in the rat 95% of optic nerve fibres cross). Crossed and uncrossed fibres from both eyes make their synaptic connections with geniculate neurons, in alternating eye-specific layers that are strictly mononocular. Geniculate neurons send their projections mainly to the layer IV of the primary visual cortex. (B) In newborn mammals axon terminals of geniculo-cortical fibres are still intermixed with each other in the layer IV of the primary visual cortex. During the critical period (in the rat from P15, time of eye opening, to P45) the visual cortex becomes subdivided into ocular dominance columns. This process is dramatically dependent upon visual experience. Indeed, if the animal is deprived of vision from one eye (monocular deprivation) the cortical territories occupied by dLGN axons functionally connected with the closed eye are strongly reduced, whereas those connected with the open eye expand.

cortical neurons are organised in functionally distinct eye specific domains (columns), according to their ability to respond exclusively, or predominantly, to eye specific visual stimuli (luminous bars) that have specific spatial orientation. These authors subdivided cortical cells in different classes depending on their "eye preference": cells in class 1 respond exclusively to contralateral eye stimulation, cells in class 4 respond equally to stimulation of both eyes and cells in class 7 exclusively to stimulation of the ipsilateral eye. Finally, cells in class 2-3 and 5-6 respond mainly to stimulation of the contra- and ipsilateral eye respectively. Today, the knowledge of the anatomical and physiological organisation of cortical neurons has improved very much, so that we can consider the cortex as an assemblage of several basically similar structures, or units, in which cortical cells are grouped according to their functional features.

Not only the anatomy, but also the functional properties of visual cortical neurons are immature during initial development. Responsive properties in the visual cortex gradually appear during the first postnatal weeks of life. Initially, cortical neurons are binocularly driven and have no orientation preference; later, most cells become monocularly dominated and acquire their orientation preference.

Experience-dependent plasticity of the visual cortex: monocular deprivation effects

Many studies have investigated the mechanisms that regulate the functional and anatomical development of the primary visual cortex. Neural activity has been indicated as the strongest force guiding these developmental changes. In particular, the visual input to the cortex plays a fundamental role in sculpting the visual system.

The main strategy used to investigate visual system organisation has been visual deprivation (Monocular Deprivation, MD). If the animal is deprived of vision in one eye for several weeks starting from the time of eye opening (postnatal day 15 in the rat) the ocular dominance distribution of neurons in the visual cortex is dramatically shifted. The great majority of cells are now monocularly driven by stimulation of the open eye, whereas the percentage of cells responding to the deprived eye decreases dramatically. At anatomical level, cellular soma of geniculate neurons connected to the deprived eye decrease in size (shrinkage); in the primary visual cortex, the territories dominated by the open eye expand, whereas those dominated by the deprived eye are relegated to very small patches (Figure 1). All these effects are not present if animals are deprived of the vision in adult age; these findings indicate the existence of the so-called "critical period", during which the development and maintenance of connections in the visual system is susceptible to dramatic alterations by abnormal visual experience.

As a consequence of their work, D.H. Hubel and T.N. Wiesel concluded that neural activity is necessary for the correct formation of geniculo-cortical connections. Their hypothesis was that, during the critical period, the formation of cortical connections is regulated by a use-dependent synaptic competition between afferent fibres from the two eyes. Many other scientists have contributed to the study of the developmental plasticity of the visual cortex (Katz and Shatz, 1996). In summary, the main findings are:

1) The timing and the patterning of neural afferent activity are necessary for a correct segregation of geniculo-cortical afferents.

2) The coincident activation of both pre- and postsynaptic neurons plays a critical role in the correct formation of the visual cortex.

3) A physiological competition between inputs from the two eyes is required for ocular segregation (binocular competition).

Neurotrophic hypothesis for the plasticity of the visual cortex

In more recent years, L. Maffei and collaborators postulated that, during the formation of geniculo-cortical connections, afferent fibres from the dLGN compete for a Neurotrophic Factor of the Nerve Growth Factor (NGF) family, produced in limited amount by target cortical cells (simply depicted in Figure 2). The production and release of this factor, and possibly its uptake, might be dependent upon afferent electrical activity. In monocularly deprived animals, electrical activity in the deprived fibres would be insufficient or inappropriate for the necessary production and/or

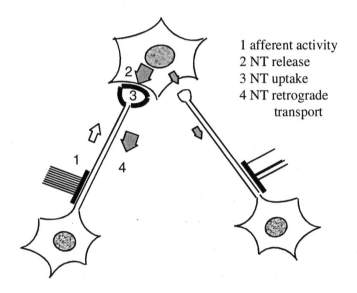

1 afferent activity
2 NT release
3 NT uptake
4 NT retrograde
    transport

Figure 2. Neurotrophic hypothesis: schematic model of the possible action of neurotrophins in the activity-dependent stabilisation of synapses. The figure represents two neurons competing for postsynaptic territory. The neuron on the left is more active than the one on the right and it is therefore able to induce a strong release of neurotrophins from the postsynaptic cell. The released neurotrophin feeds back onto the presynaptic fibre where it stabilises the synapse following its receptor-mediated uptake. Successively, the ligand-receptor complex is retrogradely transported to the neuron soma. Synaptic terminals belonging to the less active neuron receive little neurotrophin becoming atrophic and being eventually eliminated. (1) afferent activity. (2) Neurotrophin release. (3) Neurotrophin uptake. (4) Neurotrophin retrograde transport.

uptake of the neurotrophin. As a consequence, deprived fibres with insufficient Neurotrophic Factor would loose the competition with the non-deprived eye, unless Neurotrophic Factor is provided exogenously in appropriate amount. This hypothesis was confirmed by experimental analysis. They found that exogenous administration of NGF to the visual cortex totally prevents the physiological and anatomical effects of MD in rats and cats (Maffei *et al.*, 1992; Domenici *et al.*, 1991; Carmignoto *et al.*, 1993; Fiorentini *et al.*, 1995). The effects observed following the elimination of endogenous NGF supported the hypothesis of a target derived neurotrophic retrograde message, necessary for the correct development of the visual cortex. Indeed, infusing antibodies against NGF during development dramatically affected the correct development of the visual system and prolonged the critical period for the plasticity of the visual cortex (Berardi *et al.*, 1994; Domenici *et al.*, 1994). Figure 3 summarises the results of these experiments.

Neurotrophic Factors and plasticity of the visual cortex

The work of L. Maffei and colleagues opened the avenue for the study of neurotrophin involvement in cortical plasticity by other groups. The group of C.J. Shatz showed that intracortical administration of Brain-Derived Neurotrophic Factor (BDNF) and Neurotrophin-4 (NT-4) disrupts ocular dominance distribution formation in kittens (Cabelli *et al.*, 1995a; Cabelli *et al.*, 1997). L.C. Katz and colleagues showed that injections of microbeads

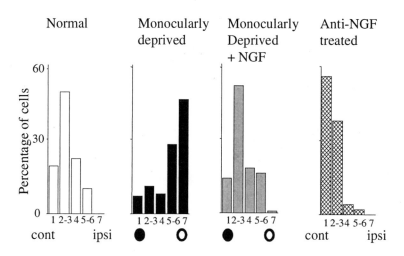

Figure 3. Effects of monocular deprivation (MD), NGF treatment and anti-NGF antibodies infusion on ocular dominance distribution in the rat visual cortex. Ocular dominance distribution of normal (white bars), monocularly deprived (black bars), NGF treated + monocularly deprived (grey bars), anti-NGF antibodies treated rats (cross-hatched bars). In the visual cortex of a normal adult rat, the majority of cortical neurons is functionally connected with the contralateral eye. Monocular deprivation during the critical period causes a marked shift of this distribution in favour of the open eye (open circle), whereas MD in adult animals has no effect. Exogenous administration of NGF in monocularly deprived rats prevents completely the shift in ocular dominance distribution caused by MD. Conversely, sequestering endogenous NGF by the transplant of anti-NGF antibodies affects the normal development of ocular dominance distribution; this treatment causes a marked loss of binocular neurons with a prominent shift toward the contralateral eye. Closed circle = deprived eye, open circle = open eye. Ocular dominance of cortical cells is classified according to D.H. Hubel and T.N. Wiesel. Ocular dominance classes: 1 and 7 respond exclusively to stimulation of the contra- and ipsilateral eye respectively; 2-3 are mainly dominated by the contralateral eye, 5-6 by the ipsilateral eye; 4 respond equally to both eyes.

conjugated to NT-4, but not to BDNF, were effective in preventing the shrinkage of ocular dominance distribution of cortical neurons induced by MD in cats (Riddle *et al.*, 1995). More recently, a clarifying work on the effects of neurotrophins on rat visual cortex plasticity was performed in L. Maffei's laboratory (Berardi *et al.*, 1998). It was indeed important to compare the actions of all four neurotrophins on the same single species, on animals of the same age and in strictly the same experimental conditions. The main findings of this work are that NGF and NT-4 are able to prevent MD effects. BDNF preventing effects were found to be clearly different from those of NGF and NT-4. Indeed, BDNF prevented MD effects only at high doses; moreover, it strongly affected the electrical activity of visual cortical neurons and reduced cell responsiveness. Neurotrophin-3 (NT-3) was found not to be effective in preventing ocular dominance shift. In another series of experiments, the group of L. Katz had demonstrated that administration of neurotrophins to visual cortical slices of neonatal ferret could induce a morphological reorganisation of synaptic arborisation, with each neurotrophin acting on specific cortical layers; this action is ineffective in absence of spontaneous electrical activity (McAllister *et al.*, 1995; 1996; 1997). More recently, L. Maffei's group produced two important works. In the first, Pizzorusso *et al.* (1999) investigated the relative roles played by the two NGF receptors, TrkA and p75^{NTR}, in mediating NGF preventing effects in MD. It was demonstrated that NGF effects are mediated mainly via TrkA high-affinity receptor, with p75^{NTR} low affinity receptor having a minor facilitator role. These results suggested the presence of specific NGF receptors at cortical level. Then, Caleo *et al.*, (1999) demonstrated that NGF signalling via TrkA receptor must be coupled to afferent electrical activity in order to produce its effects on eye preference of cortical neurons. These results are the first *in vivo* evidence of a reciprocal interaction between afferent activity and Neurotrophic Factors in modulating cortical plasticity.

All these results from the groups of L. Maffei, C.J. Shatz and L.C. Katz clearly demonstrate that Neurotrophic Factors of the NGF family, along with afferent electrical activity, play a fundamental role in the developmental plasticity of the mammalian visual system.

Modulators of cortical plasticity: the cholinergic input

Other factors that have been reported to fundamentally influence cortical development and plasticity are the cortical glutamate NMDA receptors, and the afferent systems: cholinergic, noradrenergic and serotonergic (Bear *et al.*, 1990; Kasamatsu and Pettigrew, 1979; Bear and Singer, 1986; Gu and Singer, 1993; Roerig and Katz, 1997; Wang *et al.*, 1997). The three afferent systems send a diffuse projection to the visual cortex and play a permissive role in gating ocular dominance plasticity.

In particular, it has been shown that a lesion of the cholinergic and noradrenergic inputs to the visual cortex, during the critical period, blocks cortical plasticity in the cat (Bear and Singer,

1986). The lesion of only one of the two systems left cortical plasticity unaffected. The authors postulated that during development, other neuromodulatory systems might compensate the absence of the cholinergic system. However, in L. Maffei's laboratory, by using a lesion approach, Siciliano *et al.* (1997) have recently shown that cholinergic system lesion during rat postnatal development transiently interferes with cortical competitive process.

By using a different experimental approach, Gu and Singer (1993) had demonstrated that pharmacological blockade of cholinergic receptors in the visual cortex of kittens is sufficient to strongly diminish neuronal plasticity, as tested by monocular deprivation.

Different results, then, have been obtained depending on the experimental procedure used. It must be born in mind that, contrary to pharmacological blockade procedures, that are more selective and probably more efficient, lesion procedures can affect also many different aspects of cortical physiology, in particular those related to cortical development. All together, these results indicate the cholinergic input as the major neuromodulator of cortical plasticity.

Neurotrophic Factors of the Nerve Growth Factor (NGF) family

Nerve Growth Factor is the progenitor of the Neurotrophic Factor family of neurotrophins (Cohen, 1960). These molecules are a class of highly homologue small proteins that regulate the survival and differentiation of specific classes of neurons both in the peripheral and in the central nervous system (Levi-Montalcini, 1987). The neurotrophins in mammals are NGF, BDNF, NT-3 and NT-4 (Lewin and Barde, 1996). Usually expressed in low amounts in different areas, neurotrophins have been characterised as secretory protein. Their biological action is mediated through the interaction with specific membrane receptors: the high-affinity Trk receptors belonging to the tyrosine kinase receptor family, and the low-affinity p75^{NTR} of the Tumour Necrosis Factor (TNF) receptor family. NGF binds TrkA, BDNF and NT-4 bind the same receptor TrkB, and NT-3 binds mainly TrkC. p75^{NTR} is capable of binding all those neurotrophins (Bothwell, 1995). A schematic representation of Neurotrophic Factors and corresponding receptors is reported in Figure 4.

It is generally thought that Trk receptors (molecular weight 140-145 kD) are the functionally transducing receptors for neurotrophin signalling. Ligand binding induces homodimerisation of Trk receptors, auto-phosphorylation activity and successive activation of kinase activity that results in activation of intracellular pathways involving several different kinasic proteins. As a consequence, the ligand/receptor complex is internalised and transported to the nucleus where it can exert other long-lasting actions. Truncated forms of Trk receptors, that lack the intracellular kinasic domain, have been described. Some possible functions of these forms have been postulated: they might reduce the responsiveness to neurotrophins by acting as negative mediators or

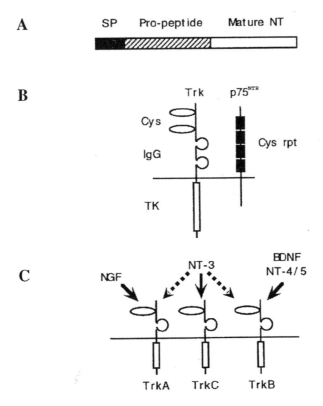

Figure 4. Neurotrophic Factors and corresponding receptors. (A) Neurotrophins (NTs) are synthesized as immature pre-pro-peptides and then processed by proteolytic cleavage in order to give the mature, biologically active form. Black box = signal peptide (SP). (B) Trk and p75[NTR] receptors. All Trk receptors (A, B and C) have a common structure. Cys, cysteine rich domains; IgG, immunoglobulin-like domain, TK, tyrosine-kinase domain; Cys rpt, cysteine repeat. (C) Binding interactions between neurotrophins and Trk receptors.

as non-functional receptors; they might have adhesive properties due to their partial homology to cell adhesion molecules; they might present ligands to full-length receptors (Barbacid, 1994; Chao and Hempstead, 1995).

p75^{NTR}, originally named low affinity NGF receptor, was the first neurotrophin receptor identified at molecular level (Chao, 1994). The extracellular domain of this protein, 75 kD in molecular weight, contains a cysteine-rich domain responsible for ligand binding. Unlike other growth factor receptors, p75^{NTR} does not have a kinasic activity in the cytoplasmic domain. Many hypotheses about its still not clear role are under investigation: p75^{NTR} may modulate Trk function either positively or negatively depending on the cellular context (inducing a direct – through formation of heterodimer – or indirect conformational change in Trk receptors); it may increase the local concentration of ligand to be presented to Trk receptors; it may play a facilitator role in the retrograde transport of the ligand/receptor complex and then in the retrograde signalling; it may act independently of Trk receptors by its own signalling cascade (sphyngomyelinase ceramide pathway); or it may act as mediator of apoptotic cell death.

Distribution of Neurotrophic Factors and corresponding receptors in the visual system

The analysis of the distribution and regulation of the expression of neurotrophins and corresponding receptors has contributed to the study of neurotrophins involvement in visual cortical plasticity and in particular helped in identifying their site of action. These results are summarised in Table 1.

NGF, TrkA, p75^{NTR}

NGF is present both at mRNA and at protein level in the rat visual cortex and reaches a peak of expression about P21 (Large *et al.*, 1986). No clear description of the laminar distribution of NGF mRNA has been reported; NGF-immunoreactivity seems to be homogenously distributed across all layers and some pyramidal neurons appear NGF-immunolabelled (Nishio *et al.*, 1994). NGF is also present in the dLGN (Schoups *et al.*, 1995), where it can exert its action on retinal ganglion cells that express NGF receptors, TrkA and p75^{NTR} (Carmignoto *et al.*, 1991).

The presence of TrkA in the visual cortex is still debated. Some authors showed that TrkA mRNA is present in the visual cortex (Miranda *et al.*, 1993; Valenzuela *et al.*, 1993; Cellerino *et al.*, 1996) whereas others failed to detect its expression in this area (Merlio *et al.*, 1992; Gibbs and Pfaff, 1994; Schoups *et al.*, 1995). Concerning the protein, some groups showed, by immunohistochemical analysis, a positive signal in few cortical areas, but not in the occipital cortex (Sobreviela *et al.*, 1994; Prakash *et al.*, 1996; Mufson *et al.*, 1997). It has, then, still to be determined whether or not this receptor is expressed at cortical level, and, if it is, which is its cellular localisation. No signal of TrkA expression has been detected in the LGN (Schoups *et al.*, 1995).

		dLGN	VISUAL CORTEX	
NGF	mRNA	+	+	**Peak at P21**
	protein	ND	+	**(all layers)**
BDNF	mRNA	?	+++	**Peak CP** **(layers II-III**
	protein	-	+++	**and V-VI)**
NT-4	mRNA	+	+	**Predominant**
	protein	ND	+	**in layer IV**
NT-3	mRNA	+	+	**Declines** **postnatally**
	protein	ND	+	**(layer IV)**
TrkA	mRNA	-	?	**Full length/truncated** **decreases in CP**
	protein	-	?	**(all layers)**
TrkB	mRNA	++	++	
	protein	++	++	
TrkC	mRNA	+	+	
	protein	+	+	
p75^{NTR}	mRNA	-	-	**Cholinergic**
	protein	?	+	**terminals?**

Table 1. Expression of Neurotrophic Factors and corresponding receptors in the mammalian geniculo-cortical system. Abbreviations used in the table: -, not present; +/++/+++, low, medium and high level of expression (arbitrary units); ND, not determined; ? contrasting results.

p75^{NTR} protein has been detected at cortical level, and its spatial distribution parallels that of cholinergic afferent terminals, suggesting the presence of p75^{NTR} on these fibres (Pioro and Cuello, 1990). Some controversy exists about the detection of p75^{NTR} protein at LGN level. It is present in this nucleus (Yan and Johnson, 1988) with the major fraction probably located on retinal ganglion cell fibres; however, Allendoerfer *et al.* (1994) failed to immunoprecipitate p75^{NTR} protein from the ferret LGN. No clear data are present to date on the expression of p75^{NTR} mRNA in the geniculo-cortical system.

BDNF, NT-4, TrkB

BDNF levels of expression in the cortex are much higher with respect to NGF. In the visual cortex BDNF is present both at mRNA and protein level with a peculiar laminar distribution (layers II-III and V-VI, with very few cells in layer IV, the main target of geniculate terminals) and a particular developmental regulation (BDNF expression increases in concomitance with eye opening, P15) (Castrén *et al.*, 1992; Bozzi *et al.*, 1995; Rossi *et al.*, 1999). No BDNF-immunoreactivity has been detected in the LGN, and there is also some controversy concerning the presence of its mRNA in this area (Schoups *et al.*, 1995; Castrén *et al.*, 1992; Conner *et al.*, 1997).

TrkB receptors are present both in the visual cortex and in the LGN and are developmentally regulated: it has been observed that while full-length receptors are predominant during early development, the level of truncated receptors dramatically

13

increases during the critical period (Allendoerfer *et al.*, 1994). At cortical level, the majority of trkB expressing neurons are of clear pyramidal morphology, but a significant number of GABAergic cells have been observed expressing this receptor (Cellerino *et al.*, 1996).

Concerning NT-4 expression, it has been found that its mRNA and protein are present in all layers of the visual cortex during the critical period, but predominantly in the layer IV (Friedman *et al.*, 1998); NT-4 mRNA has been detected also in the LGN (Cabelli *et al.*, 1995b).

NT-3, TrkC

In the rat, NT-3 mRNA is highly expressed around birth in both visual cortex and dLGN; it declines to undetectable levels during the first postnatal week of life (Schoups *et al.,* 1995). Its expression seems to be prominent in layer IV (Lein *et al.*, 1995). Also NT-3 protein has been detected at neocortical level (Friedman *et al.*, 1998)

TrkC mRNA and protein have been detected in the visual cortex and in the LGN during development (Schoups *et al.*, 1995).

Neurotrophic Factor expression: electrical activity dependence

Another important feature that has been extensively studied by different groups is the dependence of neurotrophin expression upon electrical activity. Indeed, it has been hypothesised that for neurotrophins to act as mediators of activity-dependent plasticity, their expression must be dependent upon it.

Initial studies focused on hippocampus and cerebral cortex. It has been demonstrated that increasing electrical activity results in up-regulation of NGF and BDNF, whereas the decrease of electrical activity induces a down-regulation of these factors. This regulation is mediated by classical neurotransmitters: up-regulation by glutamate via NMDA and non-NMDA receptors, and also by acetylcholine via muscarinic receptors, whereas down-regulation is predominantly mediated by GABA via GABA receptors (Zafra *et al.*, 1990; Zafra *et al.*, 1991; Berzaghi *et al.*, 1993). This regulation functions not only under extreme experimental conditions, but also in maintaining the normal physiological amount of neurotrophins. Indeed, concerning the visual cortex, Castrén *et al.* (1992) reported that also visual stimulation can modulate the levels of BDNF and TrkB mRNA in the visual cortex: rearing animals in complete darkness resulted in down-regulation of these molecules, exposure to normal light restored the normal level of BDNF expression. In L. Maffei's laboratory, we have contributed to the investigation of the role of BDNF in activity-dependent plasticity, by investigating the expression of BDNF mRNA and protein in the rat visual cortex (Bozzi *et al.*, 1995; Rossi *et al.*, 1999). The main results of these studies are that cortical BDNF expression is up-regulated by pharmacologically increasing cortical activity, and down-regulated by decreasing electrical activity afferent to the visual cortex. We also observed a developmental up-regulation of BDNF expression in coincidence with eye opening. Finally, we demonstrated that the closing of one eye, by monocular deprivation, results in a down-regulation of BDNF expression in the cortex connected to the closed eye. All these results contributed to indicate that BDNF plays a fundamental role in the plasticity of the rat visual cortex.

Neurotrophic Factor release

As already mentioned, Neurotrophic Factors have been initially characterised as secretory proteins. H. Thoenen and collaborators were the first to investigate the release of Neurotrophic Factors from neurons. These authors showed that NGF is released along two different pathways: the constitutive release, independent of electrical activity, predominantly from the surface of the cell body, and the regulated release, induced by depolarisation, from dendrites (Blöchl and Thoenen, 1995). Similar results were found to be valid also for the release of BDNF (Griesbeck *et al.*, 1995). It is to be noted that the activity-dependent release of neurotrophins seems to be regulated by mechanisms different from those used for neuropeptides and neurotransmitters, indeed it is independent of extracellular Ca^{2+} but depends upon intracellular Ca^{2+} stores and upon extracellular Na^+. More recently, Canossa *et al.* (1997) have shown that neurotrophin release can also be induced by neurotrophins themselves. The observation that neurotrophin release is spatially restricted and depends upon electrical activity, suggests that this mechanism can be involved in neurotrophic action on brain plasticity. The neurotrophin-induced neurotrophin release suggests the presence of more complex synergistic effects between different neurotrophins.

Possible mechanisms of Neurotrophic Factor action in visual cortex plasticity

Neurotrophins have been initially considered in respect to their involvement in the developing embryonic brain and in the maintenance of specific functions of certain neurons during adulthood. More recently, evidence has emerged that these factors are also involved in the process of activity-dependent developmental plasticity. A possible mechanism of neurotrophin action on neuronal plasticity is the modulation of synaptic efficacy. The first evidence for such a role came from Lohof et al. (1993), who demonstrated that Neurotrophic Factors could potentiate synaptic currents in *Xenopus laevis* neuromuscular junction, probably by increasing neurotransmitter release from presynaptic terminals. Today, many electrophysiological studies indicate that neurotrophins can modulate synaptic transmission also in the CNS. Concerning the visual cortex, it has been shown that Neurotrophic Factors can potentiate the excitatory synaptic transmission (Carmignoto et al., 1997; Akaneya et al., 1997). In a recent work, by using an *in vitro* synaptosomal preparation, L. Maffei and collaborators demonstrated that NGF and BDNF could modulate presynaptic release of neurotransmitters in the rat visual cortex during the critical period for plasticity (Sala et al., 1998). In particular, NGF potentiates the K^+-evoked release of acetylcholine and glutamate, but not of GABA; BDNF enhances the release of all the three neurotransmitters tested. This observation is particularly relevant to the understanding of the mechanism of neurotrophin action on cortical plasticity. In monocularly deprived animals, the neurotrophic action on synaptic transmission could prevent the decrease in synaptic efficacy of deprived fibres by modulating neurotransmitter release, preserving, then, the normal organisation of synaptic connections in the visual cortex.

Possible sites of Neurotrophic Factor action in visual cortex plasticity

Neurotrophic Factors play different roles in regulating visual cortical development and plasticity probably through different mechanisms and on different neuronal targets, such as LGN afferents, intracortical circuitry and subcortical afferents. Conclusions regarding NGF are reported in the Discussion.

BDNF

BDNF is present in the visual cortex during the whole postnatal life; moreover, its expression is strictly modulated by afferent electrical activity, indicating that it is involved in cortical plasticity. Its specific receptor TrkB is expressed on cortical neurons, on LGN terminals and on BFCN projecting to the cortex (Yan *et al.* 1997). All these structures are possible targets of BDNF action. Peculiar of BDNF action is the modulation of the inhibitory circuitry (Huang *et al.* 1998). TrkB is expressed also on inhibitory

18

interneurons; BDNF enhances GABA release and regulates the expression of inhibitory interneurons markers, such as Neuropeptide Y (Nawa *et al.* 1994). These effects can account for the BDNF disruption of cortical columns observed by Cabelli *et al.* (1995a) in kittens. Still, it is not clear how BDNF can partially prevent MD effects. It must be mentioned that BDNF can influence serotonergic system (Celada *et al.* 1996; Mamounas *et al.* 1995), which is a potent modulator of cortical activity (Roerig and Katz, 1997) and plasticity (Cases *et al.* 1996).

NT-4

At the level of thalamic afferents, NT-4 seems to be the most effective of all neurotrophins in preventing MD effects. Few data are available to date on its distribution and action. NT-4 mRNA is present in the layer IV of the visual cortex and its receptor TrkB is expressed also at geniculate level. Recent experimental findings indicate that, although sharing the same receptor TrkB, BDNF and NT-4 may not have similar action (Minichiello *et al.*, 1998).

NT-3

No action of NT-3 on cortical plasticity has been reported to date. NT-3 and its specific receptor TrkC are expressed in the cortex during embryonic and early postnatal life. It is possible that they have a premature action on the development of the cortex.

Outline of the experimental approach

As described above, several experiments have implicated Neurotrophic Factors of the NGF family in the developmental plasticity of the visual cortex. However, neither the sites nor the mechanisms of neurotrophin action are clearly understood.

In order to contribute to the understanding of this topic, I focused my attention on the progenitor of Neurotrophic Factors, that is Nerve Growth Factor. As the sites of a protein action can be identified from the distribution of its specific receptors, I analysed the expression of NGF receptors, TrkA and p75^{NTR}, at cortical level by using a recently produced, and particularly specific, anti-TrkA antibody (RTA; Clary *et al.*, 1994), and the anti-p75^{NTR} antibody (Promega). Since it is known that NGF action on cortical plasticity takes place mainly through interaction with TrkA receptor, particular attention was given to the study of this receptor.

Once identified the presence of TrkA receptor in the rat visual cortex, I have tried to modulate its expression during the critical period for plasticity by using different experimental protocols. In particular I analysed whether cortical NGF receptor expression is modulated by altering the visual input, by administrating exogenous NGF to the visual cortex and by lesioning the cholinergic afferences to the visual cortex, that are supposed to express such receptors.

Finally, in order to verify the possible interaction between the cholinergic neuromodulatory input to the visual cortex and the NGF effects on cortical synaptic transmission, in collaboration with the Section of Pharmacology and Toxicology, Dept. of Experimental Medicine of Genova, we measured NGF effects on neurotransmitter release in cortical synaptosomes prepared from cholinergic system lesioned animals.

RESULTS

RESULTS

TrkA receptor protein expression in the rat visual cortex

Immunohistochemistry

The expression of TrkA protein in the rat visual cortex was investigated first by using immunohistochemical techniques with RTA polyclonal antibody. No RTA-immunolabelled structures were detected in the adult rat visual cortex (Fig. 5A) with this technique; the staining observed in this area was similar to the background staining observed in sections processed without primary antibody (control sections, Fig. 5B). On the contrary, this antibody clearly stained many other structures in the adult rat CNS. In particular, as shown in Figure 5C, basal forebrain neurons were strongly immunopositive. This staining was completely eliminated by omission of the primary antibody (Fig. 5D), indicating that it represents specific labelling of endogenous TrkA receptor and not an artefact. Figure 5E is a magnification of the basal forebrain area labelled with RTA antibody.

Western blot analysis

The expression of TrkA receptor in the adult rat visual cortex was successively investigated by using western blot analysis. As shown in Figure 6, two strong bands, of the expected molecular weight of 140 and 110 kD, were evidentiated by RTA western blot in this area. PC12 cells, used as positive control, showed a positive signal, whereas epidermal rat fibroblasts (ERF), used as negative control, did not show any signal. In the basal

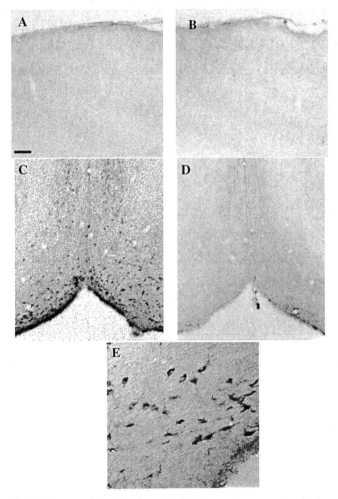

Figure 5. RTA-immunohistochemistry for TrkA receptor expression in the adult rat brain. A, B: visual cortex. (A) No RTA-immunoreactivity was observed in the adult rat visual cortex. (B) Control section of the visual cortex processed without primary antibody; the background staining is similar to the staining observed in Fig. 1A. C, D, E: basal forebrain. (C) In the basal forebrain area many neurons, corresponding to BFCN, were RTA-immunolabelled. (D) In control section of the basal forebrain area, processed without primary antibody, only background staining was observed. (E) Magnification of the basal forebrain area labelled with RTA antibody. Scale bar = 125 μm for (A) and (B), 100 μm for (C) and (D), 50 μm for (E).

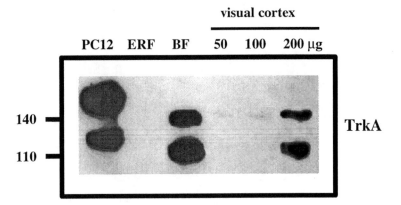

Figure 6. RTA western blot analysis for TrkA receptor expression in the adult rat visual cortex. Total protein extracts were prepared from Pheochromocytoma cells (PC12, positive control; 20 µg/lane), epidermal rat fibroblasts (ERF, negative control; 20 µg/lane), basal forebrain (BF, 50 µg/lane) and visual cortex (50, 100 or 200 µg/lane). In the adult rat visual cortex the signal was detected only with 200 µg of total protein, but not with smaller amounts. The 140 kD band corresponds to the functional form of TrkA receptor; the 110 kD band possibly to an underglycosilated form of TrkA.

forebrain a signal was already detectable by only loading on the gel 50 μg of total protein extracts/lane. On the contrary, in the adult rat visual cortex the signal was detected only with 200 μg of total protein, and not with smaller amounts. The molecular weight of the bands detected by RTA in the CNS areas investigated appeared slightly smaller than the one observed in PC12 cells, as already reported by other authors (Holtzmann *et al.*, 1995).

Specificity of RTA signal

The specificity of the signal detected by RTA in western blot analysis of visual cortex extracts was further investigated by means of immunoprecipitation tests. One milligram of protein extracts from adult rat visual cortex was immunoprecipitated with anti-TrkA antibody (Biolabs); the resulting sample was loaded on the gel and then probed with RTA for western blot analysis. As shown in Figure 7A, two bands of 140 and 110 kD in molecular weight were detected in this sample. The signal obtained is very similar to the one obtained in the non-immunoprecipitated sample. Omission of the anti-TrkA antibody (Biolabs) in the immunoprecipitating procedure completely eliminated the signal. This result suggests that the signal detected by RTA in visual cortex extracts is specific for TrkA antigen. A similar result was obtained by using RTA as the immunoprecipitating antibody and anti-TrkA (Biolabs) as the revealing antibody (data not shown).

In another set of experiments I have used the Fab fragment of RTA antibody for western blot analysis. As shown in Figure 7B,

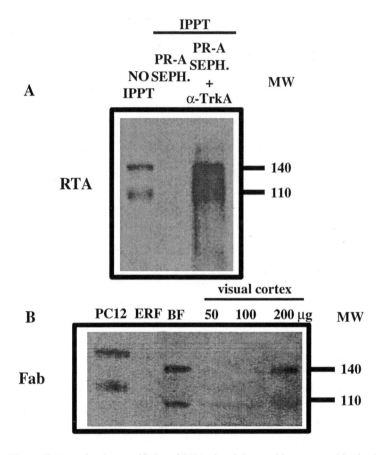

Figure 7. Tests for the specificity of RTA signal detected by western blot in the adult rat visual cortex. (A) 1mg of total protein extracts from visual cortex was immunoprecipitated with anti-TrkA antibody (Biolabs) and collected with Protein-A Sepharose. The resulting sample was loaded on a 10% poliacrylamide gel and immunoblotted with RTA. The non immunoprecipitated sample (NO IPPT), showed a normal signal. The sample subjected to immunoprecipitation only with Protein-A Sepharose but not with the anti-TrkA (Biolabs) immunoprecipitating antibody (PR-A SEPH.), presented no RTA-immunopositive signals. The immunoprecipitated sample (PR-A SEPH. + α-TrkA) showed a signal similar to the normal one: two bands of the approximate molecular weight of 140 and 110 kD. (B) Fab fragment of RTA antibody was used as the primary antibody in the western blot analysis. The signal obtained in all samples is similar to that obtained by using RTA antibody. Same abbreviations as in Fig. 2.

the number and the molecular weight of the bands detected with Fab fragment in all samples analysed, visual cortex included, are similar to those obtained with RTA antibody (see Fig. 6 for comparison).

p75^{NTR} protein expression in the rat visual cortex

Western blot analysis

In order to investigate p75^{NTR} expression in the rat visual cortex I have used the anti-p75^{NTR} antibody (Promega) for western blot analysis. A band of the approximate molecular weight of 75 kD was detected in the adult rat visual cortex, when 200 μg of total protein extracts were loaded (Fig. 8). In the basal forebrain area the signal was already detected on loading a smaller amount of protein extracts. An almost identical band was detected in PC12 (positive controls), but not in ERF (negative controls).

Developmental expression of NGF receptors in the rat visual cortex

The developmental expression of TrkA and p75^{NTR} receptors in the rat visual cortex was analysed by using western blot technique. Protein extracts were prepared from rat visual cortex at various postnatal ages (P5, 10, 15, 20, 30 and adult).

As shown in Figure 9A, TrkA protein is present in the rat visual cortex during the whole postnatal life. It is already expressed at postnatal day 5; then, its expression increases from P10-15 and

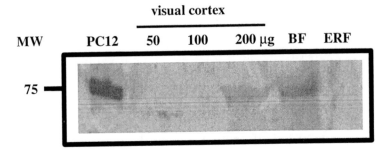

Figure 8. Western blot analysis with anti-p75^{NTR} antibody (Promega) on different samples (same as Fig. 2). A band of the approximate molecular weight of 75 kD was observed in the visual cortex only when 200 µg of total protein extracts were loaded, but not with smaller amount. A similar signal was observed also on PC12 and BF (positive controls), but not in ERF (negative control).

reaches a plateau level at P30 that is approximately maintained in the adult age.

The study for the detection of p75^{NTR} expression in the developing rat visual cortex indicated that also this protein is present at every postnatal age analysed (Fig. 9B). Unlike TrkA expression, there is not a developmental regulation for p75^{NTR} in the visual cortex.

Figure 9C shows a representative signal obtained by using anti-β-tubuline antibody (Sigma), after stripping the same filters previously probed with RTA or anti-p75^{NTR} antibody. This signal, representative of a housekeeping gene whose expression does not change during development, was used for the quantification of NGF receptors signal (see Materials and Methods).

The graph (Fig. 9D) reports the quantification of the optical density ratio between TrkA or p75^{NTR} and β-tubuline signals at the different ages analysed, normalised to the signal obtained at P5. Whereas TrkA expression increases approximately 1.5 fold from P10 to P30, approximately in concomitance with the critical period for cortical plasticity, p75^{NTR} expression does not vary during the same postnatal period of life.

Modulation of cortical NGF receptor expression
Silencing of the visual input

The dependence of cortical TrkA and p75^{NTR} expression upon afferent electrical activity was investigated by pharmacological approach. In particular, the visual input was

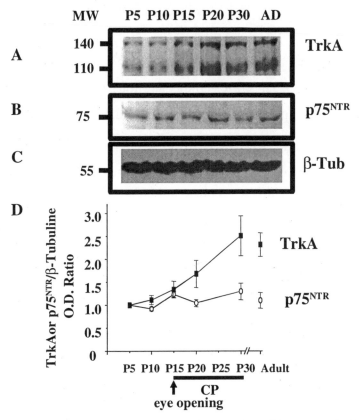

Figure 9. Developmental expression of TrkA and p75^NTR receptors in the rat visual cortex. A, B, C: western blot analysis. Visual cortex extracts were prepared from rats at different postnatal ages (P5, 10 15, 20, 30 and adult). (A) RTA-immunoblotting revealed an increase in TrkA receptor expression during postnatal life in the rat visual cortex. (B) No change in anti-p75^NTR immunolabelling was detected at the different ages analysed. (C) Filters probed with RTA or anti-p75^NTR antibodies were stripped and reprobed with anti-β-tubuline antibody (Sigma). The figure reports a representative example of the signal obtained on these filters with anti-β-tubuline antibody. No change in β-tubuline signal was observed in the visual cortex during development. (D) The graph reports the ratio between optical density values of TrkA signal (filled squares) or p75^NTR signal (open circles) and β-tubuline signal, as normalised to the data obtained at P5. Each point represents the mean±SEM of different experiments. CP, critical period. The arrow indicates the time of eye opening in the rat, P15.

silenced by means of intraocular injections of Tetrodotoxin (sodium channel blocker) in one eye. Both short-term (a single injection at P30) and chronic (one injection every 24-36 hrs for 7 days, from P23 to P30 in concomitance with the critical period) treatments were used. TTX was injected in the right eye and saline solution in the left eye as an internal control. It is to be remembered that in the rat, optic nerve fibres cross almost completely at the level of the optic chiasm (see Introduction); as a consequence, the visual cortex contralateral to the TTX-injected eye is the one that receives the silenced input.

As shown in Figure 10A, both the short term and the chronic monocular blockade of spike activity did not induce any modulation in the expression of TrkA in the visual cortex. This signal is similar to the one obtained in animals treated with saline solution and in untreated animals of the same age (P30).

A similar result was obtained in the study of p75^{NTR} expression (Fig. 10B). The short term and the chronic TTX treatment induced no modulation of cortical p75^{NTR} expression as compared to the signal obtained in the visual cortex of saline solution injected animals and in P30 untreated animals.

Figure 10C shows a representative signal obtained by using anti-β-tubuline antibody (Sigma), after stripping the same filters previously probed with RTA or anti-p75^{NTR} antibody. Optical density measurements of the signals obtained confirmed the absence of any modulation in cortical NGF receptor expression following TTX treatments (data not shown).

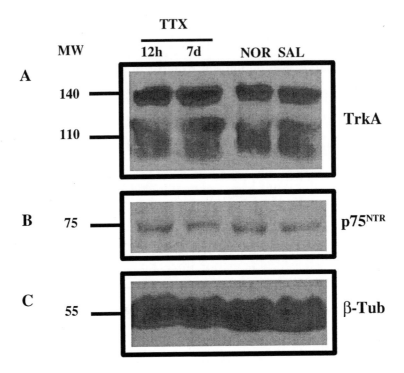

Figure 10. Modulation of cortical NGF receptor expression by visual input. Animals received intraocular injection of Tetrodotoxin (TTX) in the right eye (short treatment, single injection at P30 and tissue extraction after 12 h; chronic treatment, one injection every 24-36 hrs for 7 days, from P23 to P30, 7d). The left eye was injected with saline solution by using the same experimental procedure. No modulation in the expression of TrkA (A) and of p75^{NTR} (B) was detected in the visual cortex of TTX injected rats (neither in the short-term nor in the chronic treatment), as compared to the visual cortex of saline solution injected animals (SAL) and to what observed in normal P30 animals (NOR). (C) β-tubuline expression did not change following TTX injection. This signal was used for quantification of the results (data not shown).

Administration of exogenous NGF to the visual cortex

I investigated whether exogenous NGF administration to the visual cortex could modulate NGF receptor expression in this area. An osmotic minipump was loaded with NGF ($1\mu g/\mu l$) and implanted in the left visual cortex of young animals (P23); seven days later, the animals were sacrificed and visual cortices extracted for western blot analysis. Animals implanted with a minipump loaded with cytochrome C ($1\mu g/\mu l$) were used as controls.

As shown in Figure 11A and B, the expression of TrkA and $p75^{NTR}$ receptors in the left visual cortex (NGF-treated side) is similar to that in the right visual cortex (cytochrome C-treated side); the signal detected in treated animals does not differ from the one obtained in untreated ones. As shown in Figure 11C, NGF treatment did not induce any modulation in the β-tubuline expression, which was used for quantification of the results (data not shown).

Lesion of basal forebrain cholinergic neurons that selectively project to the visual cortex

It is known that cholinergic basal forebrain nuclei (BFCN) express high levels of NGF receptors and send a diffuse projection to the neocortex, including the visual cortex (Carey and Rieck, 1987; Dinopoulos *et al.*, 1989; Sobreviela *et al.*, 1994; Li *et al.*, 1995). I investigated whether the NGF receptors signal detected in the rat visual cortex in my experiments was due to the presence of

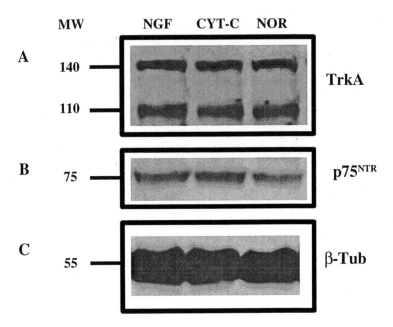

Figure 11. Modulation of cortical NGF receptor expression by NGF administration. Osmotic minipumps were loaded with NGF or cytochrome C (1µg/µl) and implanted in the left or right visual cortex of young animals (P23) respectively; seven days later animals were sacrificed and visual cortices extracted for western blot analysis. No modulation in the expression of TrkA (A) and of p75NTR (B) was detected in the visual cortex that received NGF administration (NGF). The signal is similar to what observed in cytochrome C treated animals (CYT-C) and in untreated ones (NOR). (C) Both NGF and cytochrome C treatment induced no modulation in β-tubuline expression, that was used for quantification of the results (data not shown).

cortical NGF receptors-bearing terminals afferent from the basal forebrain area. Therefore, I eliminated basal forebrain nuclei projecting to the visual cortex and performed western blot analysis for NGF receptors on visual cortices extracted from lesioned animals. On the model of experiments previously performed in our laboratory by Siciliano *et al.* (1997), animals were subjected to injections of quisqualic acid (a non-NMDA receptor agonist) in basal forebrain nuclei of the right side of the brain, that send their projections ipsilaterally to the right visual cortex (the horizontal limb of the diagonal band of Broca - hdbB - and the nucleus basalis Magnocellularis -nbM). In a couple of weeks, this treatment induces degeneration of cells and corresponding projections. Visual cortices of lesioned animals were used for verifying the efficacy of the lesion and for detecting NGF receptor expression by western blot analysis; PBS-injected and untreated animals were used as controls. Because of the high doses of quisqualic acid used (from 0.5 to 1 µl per injection site), and probably because of drug diffusion, all lesion effects analysed, as reported below, were observed mainly in the ipsilateral treated side, and, to a lesser extent, also in the contralateral side.

Stereotaxic coordinates for the drug injection

As shown in Figure 12A, the correct stereotaxic coordinates for selectively lesioning the two nuclei that specifically project to the visual cortex are 0.8 mm anterior to bregma, 0.8 mm from the midline and 7.7 mm from the pial surface for hdbB; 0.5 mm

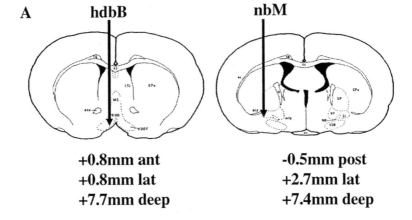

A   hdbB   nbM

+0.8mm ant   -0.5mm post
+0.8mm lat   +2.7mm lat
+7.7mm deep   +7.4mm deep

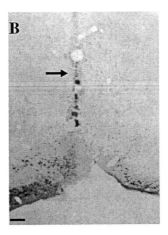

Figure 12. Experimental procedure for selectively lesioning basal forebrain nuclei that specifically project to the visual cortex. (A) Stereotaxic coordinates are: 0.8 mm anterior to bregma, 0.8 mm from the midline and 7.7 mm from the pial surface for horizontal limb of the diagonal band of Broca (hdbB); 0.5 mm posterior to bregma, 2.7 mm from the midline and 7.4 mm from the pial surface for nucleus basalis Magnocellularis (nbM). Arrows indicate micropipette penetration into the tissue. (B) ChAT-immunoreactivity on a representative section at the level of hdbB of a rat injected with blue pontamine. The injection trace is clearly visible (indicated by the arrow) and terminates in an area rich in ChAT-immunolabelled neurons. Scale bar = 80 μm for (B).

posterior to bregma, 2.7 mm from the midline and 7.4 mm from the pial surface for nbM. These coordinates were previously obtained in our laboratory by Siciliano *et al.* (1997) by injecting the neuronal tracer Wheat Germ Agglutinin-Horse Radish Peroxidase conjugated (WGA-HRP) in the basal forebrain area at different coordinates with respect to bregma (see Materials and Methods).

In order to verify whether our injection procedure was correct, I injected a vital dye (blue pontamine) at those stereotaxic coordinates and performed on the same animals an immunohistochemical analysis with Choline Acetyltranferase antibody (anti-ChAT antibody, Chemicon) for detection of basal forebrain cholinergic neurons. Figure 12B shows a representative section at the level of hdbB. The injection trace is clearly visible and it terminates in an area rich in ChAT-immunolabelled neurons, indicating that the injection protocol used was correct. A similar result was obtained at the level of nbM (data not shown).

Lesion effects on cholinergic markers

I have analysed the effect of quisqualic acid injections on different cholinergic markers, both at basal forebrain and cortical level. ChAT-immunohistochemistry was used as a marker for the effect of the lesion on the basal forebrain (this antibody stains cholinergic neurons). Figure 13A shows the normal ChAT-immunoreactivity in a quisqualic acid-injected animal, at the level of the nbM contralateral to the injected side. As reported in Figure 13B, the lesion induced a strong decrease of ChAT-containing neurons in the nbM ipsilateral to the injection. Figure 13C is a

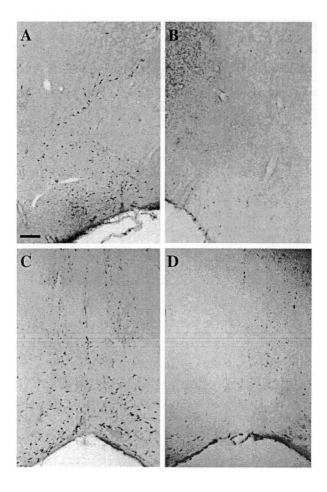

Figure 13. BFCN lesion effects on ChAT-immunoreactivity at the level of hdbB and nbM. (A) Normal ChAT-immunolabelling detected at the level of the nbM contralateral to the injected side. (B) Brain section at the level of the nbM ipsilateral to the injected side, where a clear decrease in ChAT-immunoreactivity is visible. (C) Brain section of an untreated animal at the level of hdbB with the normal pattern of ChAT-immunolabelling. (D) The lesion induced a strong decrease of ChAT-immunoreactivity in the hdbB, not only in the ipsilateral injected side (on the left), but also in the contralateral side (on the right). No modulation in ChAT immunoreactivity was observed in PBS-injected animals (data not shown). Scale bar = 150 µm for (A) and (B), 125 µm for (C) and (D).

section from an untreated animal, at the level of hdbB immunolabelled with anti-ChAT antibody; many labelled neurons are clearly visible. As reported in Figure 13D, the lesion induced a strong decrease of ChAT-immunolabelled neurons also in the hdbB; probably due to drug diffusion, this effect was clearly present also in the contralateral side. No modulation in ChAT immunoreactivity was observed in PBS-injected animals (data not shown).

Acetylcholinesterase histochemistry on brain sections of treated and untreated animals was used to detect the effect of the lesion at cortical level. As shown in Figure 14, quisqualic acid injection resulted in a marked loss of AchE staining mainly in the visual cortex ipsilateral to the injection site (Fig. 14B), but also in the contralateral visual cortex (Fig. 14C), as compared to untreated animals (Fig. 14A) and PBS-injected (data not shown) animals.

Choline uptake and Choline Acetyltranferase activity (ChAT-activity) were measured on visual cortex synaptosomal preparation obtained from quisqualic acid-, PBS-injected and untreated animals. As shown in Fig. 15, choline uptake was approximately reduced by about 42% in the cortex ipsilateral to the lesion as compared to untreated animals. Although to a smaller extent, there is a statistically significant reduction in choline uptake also in the contralateral visual cortex by about 31%. PBS-injected animals did not show a significant modulation of choline uptake, as compared to untreated animals.

ChAT-activity in synaptosomal preparation from lesioned animals was dramatically reduced in the ipsilateral cortex as

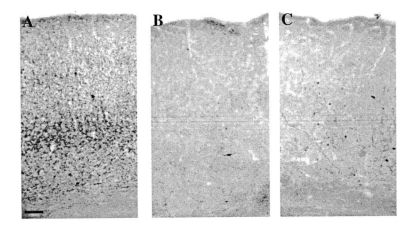

Figure 14. BFCN lesion effects on Acetylcholinesterase histochemistry in the visual cortex. (A) AchE-labelled fibres are present in all layers of the visual cortex of an untreated animal, and in particular in layers IV and V. (B) AchE-labelling strongly decreases in the visual cortex ipsilateral to the quisqualic acid injection side. (C) A clear decrease is also present in the contralateral visual cortex. Scale bar = 90 μm.

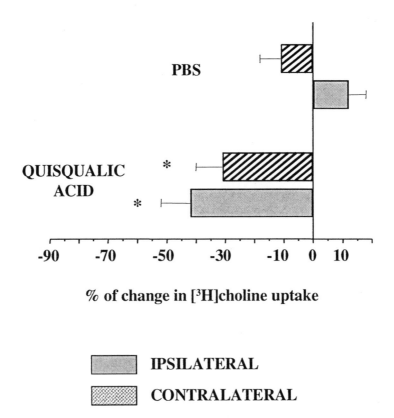

Figure 15. BFCN lesion effects on choline uptake from visual cortical synaptosomes. Each point represents the mean±SEM of different experiments run in triplicate. Results are reported as percentage of change in choline uptake in the ipsilateral and contralateral visual cortex of quisqualic acid- and PBS-injected animals as compared to untreated animals. Statistical calculations were performed using Student's two-tailed *t*-test: *P<0.05 when compared to untreated animals.

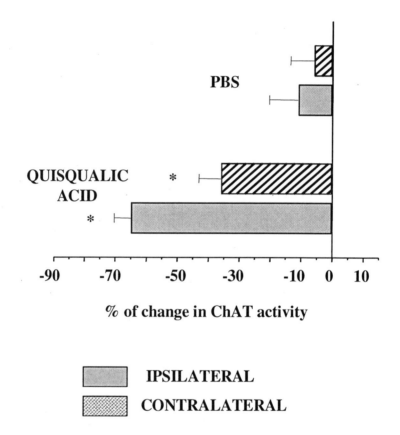

Figure 16. BFCN lesion effects on Choline acetyltranferase activity (ChAT-activity) from visual cortical synaptosomes. Each point represents the mean±SEM of different experiments run in duplicate. Results are reported as percentage of change in ChAT-activity in the ipsilateral and contralateral visual cortex of quisqualic acid- and PBS-injected animals as compared to untreated animals. Statistical calculations were performed using one way ANOVA and post-hoc Tukey's test: *P<0.05.

compared to untreated animals. As shown in Figure 16, ChAT-activity decreases by about 65% in the visual cortex ipsilateral to the injected side; a minor but significant decrease of ChAT-activity (about 35%) is present also in the contralateral side. The treatment with PBS did not vary ChAT-activity with respect to untreated animals.

I also tried some pilot experiments which indicated that BFCN lesion obtained with a different experimental protocol, that is by using 192 IgG-saporin, decreases cortical ChAT-activity similarly to the results obtained by using quisqualic acid injections (data not shown).

Lesion effects on cortical TrkA receptor expression

Finally, the effects of basal forebrain nuclei lesion on the NGF receptor expression in the visual cortex were analysed by western blot. As shown in Figure 17A, the lesion induced a strong decrease of TrkA protein expression mainly in the ipsilateral visual cortex of quisqualic acid-treated animals, but also in the contralateral cortex. The level of TrkA expression in the ipsi- and contralateral visual cortex of PBS-injected animals is similar to that of normal untreated animals. As shown in the graph (Fig. 17B), cortical TrkA/β-tubuline optical densities ratio in the ipsilateral side of quisqualic acid treated animals, decreases approximately by about 72% in comparison to the signal obtained in untreated animals, and by about 44% in the contralateral side. No significant changes were observed in PBS-injected animals. Despite the strong

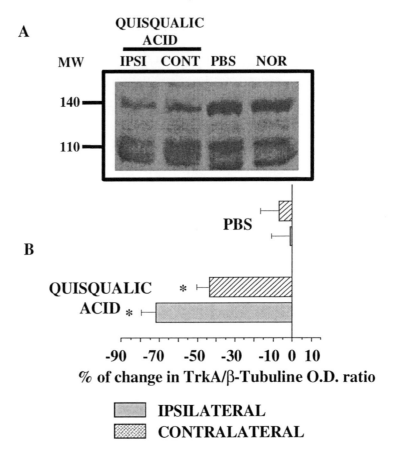

Figure 17. BFCN lesion effects on cortical TrkA protein expression. (A) Western blot representative of the decrease in cortical TrkA expression induced by BFCN lesion. In quisqualic acid-treated animals TrkA signal decreases strongly but not completely in the visual cortex ipsilateral to the injection side (IPSI). Although to a lesser extent, this signal decreases also in the contralateral visual cortex (CONT). No modulation of cortical TrkA was detected in visual cortex from PBS-injected (PBS) and from untreated animals (NOR). (B) β-tubuline signal was used for quantification of the results (data not shown). The graph reports the ratio between cortical TrkA/β-tubuline optical densities values (mean±SEM of different experiments) for the ipsilateral and contralateral visual cortex of quisqualic acid- and PBS-injected animals as compared to untreated animals. Statistical calculations were performed using one way ANOVA and post-hoc Tukey's test: *P<0.05.

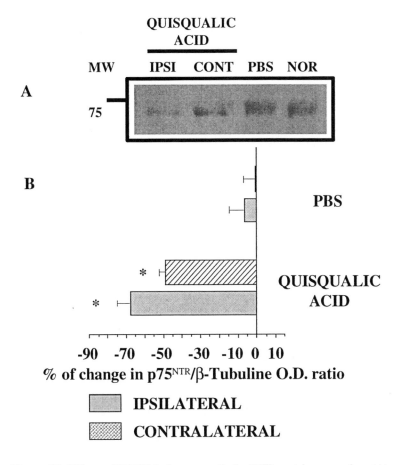

Figure 18. Effects of BFCN lesion on cortical p75^{NTR} protein expression. (A) Western blot representative of the decrease in cortical p75^{NTR} expression induced by BFCN lesion. p75^{NTR} signal decreases strongly but not completely in the visual cortex ipsilateral to the injected side (IPSI) of quisqualic acid-treated animals. Although to a lesser extent, this signal decreases also in the contralateral visual cortex (CONT). No modulation of cortical p75^{NTR} expression was detected in visual cortex from PBS-injected (PBS) and from untreated animals (NOR). (B) β-tubuline signal was used for quantification of the results (data not shown). The graph reports the ratio between cortical p75^{NTR}/β-tubuline optical densities values (mean±SEM of different experiments) for the ipsilateral and contralateral visual cortex of quisqualic acid- and PBS-injected animals as compared to untreated animals. Statistical calculations were performed using one way ANOVA and post-hoc Tukey's test: *$P<0.05$.

decrease, TrkA receptor expression in the visual cortex of quisqualic-acid injected animals is not completely eliminated; indeed a light positive band is still clearly visible following the lesion.

Lesion effects on cortical p75^{NTR} expression

Similar results were obtained in the study of p75^{NTR} expression in quisqualic acid-injected animals (Fig. 18). After the lesion, p75^{NTR} signal is strongly decreased in the visual cortex ipsilateral to the injected side, and to a lesser extent also in the contralateral cortex. The decrease of cortical p75^{NTR}/β-tubuline optical densities ratio in the ipsilateral side of quisqualic acid-treated animals is approximately 68%, and in the contralateral side 49% as compared to the signal obtained in untreated animals; no effects were observed after injections of PBS (Fig. 18A and B). The expression of p75^{NTR} is not totally eliminated; indeed, a light band is still clearly visible after the lesion.

NGF receptor expression and modulation in cortical synaptosomes from normal and lesioned animals

I investigated whether NGF receptor expression was detectable by western blot analysis also on the synaptosomal preparation from visual cortices of P30 normal animals. As shown in Figure 19A and B, both receptors are present in this experimental sample. As already observed on total protein extracts, the signal for TrkA and p75^{NTR} in visual cortical synaptosomes was

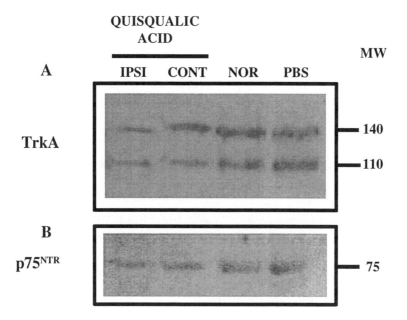

Figure 19. RTA western blot analysis for TrkA and p75^NTR protein expression in visual cortical synaptosomes prepared from normal and BFCN lesioned animals. (A) A signal similar to that obtained in total protein extracts from visual cortex of normal animals was obtained also on cortical synaptosomal preparation. Two bands, of the approximate molecular weight of 140 and 110 kD, were detected on synaptosomes prepared from normal P30 animal (NOR). BFCN lesion induced a strong decrease of cortical synaptosomes TrkA expression, mainly in the ipsilateral cortex (IPSI), but also in the contralateral one (CONT), as compared to the signal obtained in PBS-injected (PBS) and untreated animals (NOR). (B) A similar result was obtained in the analysis of p75^NTR expression on visual cortical synaptosomes. A 75 kD band was identified in western blot analysis (NOR). Following BFCN lesion, cortical p75^NTR signal decreases in the ipsilateral (IPSI) and in the contralateral (CONT) visual cortex of lesioned animals as compared to PBS-injected (PBS) and untreated animals (NOR).

detected only when 200 µg of protein extracts were used but not with smaller amounts.

Next, I analysed NGF receptor expression in synaptosomes prepared from visual cortex of quisqualic acid-, PBS-injected and untreated animals. As already observed in the total protein extracts preparation, also in the synaptosomal preparation the lesion induced a strong decrease of TrkA and p75^{NTR} receptor expression as compared to untreated and PBS-injected animals. The effects were observed mainly in the sample of the visual cortices ipsilateral to the lesioned side, but also in the contralateral ones. Also the extent of the decrease in NGF receptor expression in cortical synaptosomes following BFCN lesion was similar to the one obtained in total protein extracts (data not shown).

NGF effects on neurotransmitter release from visual cortex synaptosomes of lesioned animals

Figure 20 reports the NGF-potentiating effects on K$^+$-evoked acetylcholine release in quisqualic acid-, PBS-injected and untreated animals. In untreated animals NGF potentiates K$^+$-evoked acetylcholine release by about 35%. PBS injection does not modulate NGF potentiating effects on acetylcholine release. In quisqualic acid-treated animals NGF is no more able to potentiate glutamate release, either in the visual cortex ipsilateral to the injected side or in the contralateral cortex.

A similar result was obtained in the study of the NGF-potentiating effects on K$^+$-evoked glutamate release. Indeed, as

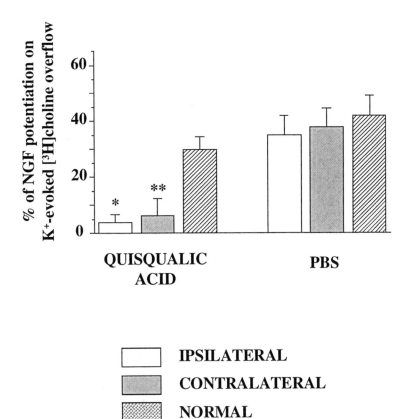

Figure 20. NGF effects on K$^+$-evoked [^{3}H]choline release from synaptosomes isolated from quisqualic acid-, PBS-injected and normal P30 rat visual cortex. NGF was used at a concentration of 100 ng/ml; KCl was 15 mM. Each point represents the mean±SEM of different experiments run in triplicate. The baseline represents the K$^+$-evoked [^{3}H]choline release without NGF. Statistical calculations were performed using Student's two-tailed t-test: *P<0.005 and **P<0.05 when compared to untreated animals.

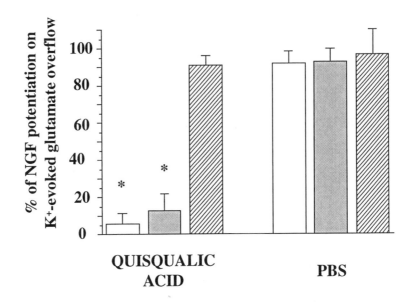

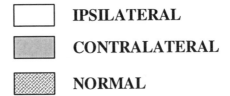

IPSILATERAL

CONTRALATERAL

NORMAL

Figure 21. NGF effects on K+-evoked glutamate release from synaptosomes isolated from quisqualic acid-, PBS-injected and normal P30 rat visual cortex. NGF was used at a concentration of 100 ng/ml; KCl was 15 mM. Each point represents the mean±SEM of different experiments run in triplicate. The baseline represents the K+-evoked glutamate release without NGF. Statistical calculations were performed using Student's two-tailed *t*-test: *P<0.0001 when compared to untreated animals.

shown in Figure 21, NGF-potentiating effects of K^+-evoked glutamate release are completely eliminated in both ipsilateral and contralateral samples from quisqualic acid-treated animals. In PBS-injected and untreated animals, NGF is still able to potentiate K^+-evoked glutamate release by about 90%.

DISCUSSION

DISCUSSION

Summary

In the present study I obtained the first direct evidence of the expression of NGF high-affinity receptor protein, TrkA, in the postnatal rat visual cortex. TrkA receptor is present at cortical level during the whole postnatal life and it is up regulated during the critical period for cortical plasticity. The developmental analysis of low-affinity NGF receptor protein, p75^{NTR}, indicated that it is present in the visual cortex but it is not modulated during postnatal development. The level of expression at cortical level of both NGF receptors, TrkA and p75^{NTR}, is modulated neither by alteration of the visual input nor by administration of NGF during the critical period. Elimination of basal forebrain cholinergic terminals afferent to the visual cortex induces a strong decrease in cortical NGF receptor expression, indicating that these terminals express NGF receptors. Many evidences suggest that cortical NGF receptors are present also in other structures, possibly cortical glutamatergic neurons, cholinergic interneurons and terminals from other afferent systems. In this study, a residual cortical NGF receptors signal was still detected following BFCN lesion. By measuring NGF potentiation of K^+-evoked neurotransmitter release, cortical NGF responsiveness was found to be dramatically impaired by BFCN lesion; this result suggests that the cholinergic input modulates the total NGF responsiveness in the rat visual cortex, possibly regulating the expression and the functionality of NGF receptors at cortical level.

TrkA receptor protein is expressed in the rat visual cortex

TrkA receptor protein expression in the visual cortex of the rat was investigated by using the specific anti-TrkA antibody RTA. By western blot analysis, I obtained the first direct evidence of the presence of TrkA receptor in the rat visual cortex during development. The levels of TrkA expression detected in the visual cortex are very low as compared to other TrkA expressing areas (i.e. basal forebrain); indeed, a large amount of total protein extracts from the visual cortex was needed in order to detect a positive signal.

Whereas RTA antibody can detect a cortical TrkA-positive signal when used for western blot analysis, no immunopositive structures in the visual cortex were revealed by immunohistochemical analysis. This latter result is in agreement with a previous work by Sobreviela *et al.* (1994) who, indeed, failed to detect TrkA receptor protein expression in the rat occipital cortex, by using the same antibody and a similar immunohistochemical protocol. These and other authors (Prakash *et al.*, 1996; Mufson *et al.*, 1997) revealed low levels of RTA-immunoreactivity in other cortical areas, such as frontal, cingulate, pyriform, enthorynal and barrel somatosensory cortex.

An explanation for the different results obtained by the two techniques could lie in the fact that they involve a different environment for antigene/antibody interaction. In the western blot analysis, proteins loaded on the gel are denaturated, so that their epitopes are easily presented to and recognised by the antibody. In the immunohistochemical technique, the interaction antigene/antibody strictly depends on different parameters, such as the penetration of the antibody in the tissue and the conditions used for tissue fixation. Moreover, a protein expressed at low levels, such as TrkA in the visual cortex, is more easily detected by western blot than by immunohistochemistry. Indeed, TrkA proteins extracted from an entire visual cortex are spatially concentrated when loaded on a gel, whereas they are distributed on many different structures in a tissue section. A TrkA signal is detected by both techniques exclusively where its level of expression is very high, that is in the basal forebrain area. Indeed, I found that many neurons were labelled in the basal forebrain area, where, as already indicated by other groups (Sobreviela *et al.*, 1994; Holtzmann *et al.*, 1995; Li *et al.*, 1995), TrkA reaches its highest level of expression in the brain.

Verification of TrkA signal detected in the rat visual cortex

The first aim was to verify the specificity of the TrkA signal obtained in the visual cortex by western blot analysis. These controls were based on the use of positive and negative control samples (cell lines and brain tissues), on the number and molecular weight of the bands detected, immunoprecipitating tests, and on the use of different anti-TrkA antibodies (including Fab fragment of RTA antibody).

Initially, I checked the specificity of the antibody in use (RTA). In our experiments, pheochromocytoma cells (PC12) - a neuronal-like cell line known to express high levels of TrkA (Kaplan *et al.*, 1991), showed a strong signal and epidermal rat fibroblasts (ERF), used as negative control, did not shown RTA-immunolabelling. As already revealed in the immunohistochemical analysis, basal forebrain presented a strong signal in the western blot, confirming the high level of TrkA expression in this area. The group that produced RTA antibody (Clary *et al.*, 1994), by using cell line transfected with plasmids containing the different Trk receptors, PC12 and sympathetic neurons, showed that RTA does not recognise TrkB or TrkC, but exclusively TrkA.

The number and the molecular weight of the bands detected in this study correspond to what already published by other authors. I evidentiated two main bands in all brain areas analysed, one of the approximate molecular weight of 140 kD and the other of 110 kD. Earlier investigations (Kaplan *et al.*, 1991; Klein *et al.,* 1991; Meakin and Shooter, 1991; Hempstead *et al.*, 1992) indicated that the 140 kD band is the mature form of TrkA that serves as the functional NGF receptor. The 110 kD form is an immature precursor of TrkA; it is also possible that the 110 kD form is a differentially glycosilated form or a breakdown product of TrkA that might arise during tissue extraction. As already shown by Holtzmann *et al.* (1995), the molecular weight of the bands detected in brain areas appears slightly smaller than the one observed in PC12 cells. Again, this could be due to a difference in the level of glycosilation in TrkA forms obtained from different preparations, as already observed by Wolf *et al.* (1995) on an insect cell line.

Also the immunoprecipitation test confirmed the specificity of this antibody. Indeed, the fact that RTA is still able to detect a positive signal on a visual cortex sample immunoprecipitated with a different anti-TrkA antibody (Biolabs), which recognises different epitopes on the same TrkA antigene, enormously decreases the probability of a possible non specific interaction between RTA and endogenous TrkA protein. A similar result was obtained by using anti-TrkA (Biolabs) blotting on RTA immunoprecipitated samples. Because the signal obtained with RTA was stronger than the one obtained with anti-TrkA antibody (Biolabs), experiments were performed exclusively by using RTA antibody.

RTA is a polyclonal antibody that binds, dimerises and activates TrkA receptor; it recognises a specific epitope (in the extracellular domain) on two TrkA receptors at the same time. Theoretically, it is possible that, because of its bivalent feature, RTA recognises some other protein linked to a TrkA receptor, or, that the free binding site residual after binding to one TrkA receptor, recognises some other protein. The results obtained with Fab fragment of RTA antibody, that has a monovalent binding site, were completely comparable to those obtained with RTA, and further confirmed the very high specificity of RTA for TrkA receptor.

p75^{NTR} protein is expressed in the rat visual cortex

Also the expression of low affinity p75^{NTR} NGF receptor in the rat visual cortex was investigated. To date few works are extant on p75^{NTR} expression in the visual cortex. In particular, no data have been obtained concerning the expression of the p75^{NTR} mRNA. Concerning the expression of p75^{NTR} protein, Pioro and Cuello (1990), by using immunohistochemistry with a different anti-p75^{NTR} antibody, evidentiated a very light and fine plexus of p75^{NTR} immunoreactive fibres in the adult rat occipital cortex. The laminar localisation of these immunopositive fibres is similar to that of cholinergic afferents to the visual cortex (layer IV and V); from these evidences they concluded that p75^{NTR} is presumably located on cholinergic terminals from basal forebrain nuclei (see below).

In agreement with Pioro and Cuello's results, by means of western blot analysis with an anti-p75^{NTR} antibody from Promega, I was able to reveal the presence of the low affinity NGF receptor p75^{NTR} in the rat adult visual cortex. Again, a large amount of total protein extracts were needed to detect a positive signal, indicating a low level of p75^{NTR} expression in this area.

Cortical TrkA and p75^{NTR} are differentially modulated during visual cortex development

Previous results obtained in L. Maffei's laboratory showed that NGF plays a key role in the developmental plasticity of the visual cortex (Maffei et al., 1992; Domenici et al., 1991) and that

40

its action is mediated mainly through cortical TrkA receptor and to a lesser extent through p75^{NTR} (Pizzorusso *et al.*, 1999). The modulation of the level of expression of a particular receptor is one of the possible strategies to regulate the efficacy of its specific ligand. In consequence, it seemed interesting to analyse whether TrkA and p75^{NTR} expression is modulated during development of the rat visual cortex.

The results obtained showed that TrkA level of expression is finely regulated during development. In particular there is a strong (2.5-fold) increase in TrkA protein expression from P10 to P30, when it reaches a level that is approximately maintained in the adult age. This lapse of time corresponds to the critical period for the plasticity of the visual cortex. It has been postulated that a gene involved in this process would be modulated specifically during the critical period, and indeed TrkA expression is up regulated in this period. In conclusion, it is possible that the up-regulation of TrkA receptor, observed in my experiments, is a step needed to allow endogenous NGF effects to take place.

Concerning the expression of cortical p75^{NTR} during development, a result similar to that obtained in the analysis of TrkA expression was expected. Indeed, it is generally thought that TrkA and p75^{NTR} are concomitantly needed for the formation of a functional NGF receptor (Chao, 1994); moreover, they often co-localise in many brain areas. In my experiments, I found no modulation of p75^{NTR} expression in the visual cortex at the different postnatal ages analysed. In their work, Sobreviela *et al.* (1994) showed that, in the basal forebrain area (where NGF receptors are abundantly expressed), most TrkA positive neurons co-localise with p75^{NTR}. However, of the basal forebrain cholinergic neurons, that send their projections to the cortex, 80% coexpress both receptors, but a 20% express p75^{NTR} alone. More recently, it has been observed that p75^{NTR} can induce its own signalling cascade independently from TrkA receptor (Carter *et al.*, 1997; Casaccia-Bonnefil *et al.*, 1996). Yeo *et al.* (1997) confirmed *in vivo* the current idea that the two receptors can also be involved in different functions; in particular, p75^{NTR} can either positively or negatively regulate neuronal features depending on the cellular context and also independently from TrkA presence.

In conclusion, regardless whether or not the two receptors co-localise at cortical level, the experiments of this study suggest that they are differentially modulated during development of the visual cortex.

Cortical TrkA and p75^{NTR} expression in the visual cortex is not dependent upon visual input

It is known that the visual input plays a fundamental role on many anatomical and physiological features of the visual system, presumably through the action on a large series of genes, including Neurotrophic Factors of the Nerve Growth Factor family. For instance, in a previous work we showed that the expression of BDNF in the visual cortex is modulated by the visual input and that this factor is involved in the developmental plasticity of the visual cortex (see Introduction). Concerning NGF its level of expression seems not to be modulated by the visual input. However, the dependence of neurotrophin action upon afferent electrical activity would be similarly obtained by modulating the expression of their specific receptors. Indeed, it has been recently shown that the expression of Trk receptors can be modulated by electrical activity in particular brain areas (Castrén *et al* 1992; Bengzon *et al.*, 1993).

In this study cortical TrkA expression was found to increase during development starting approximately around the time of eye opening (P15). It seemed interesting to analyse whether eye opening, by inducing a sudden increase of electrical input to the visual cortex, could up-regulate cortical TrkA expression. Against this hypothesis, silencing of the afferent electrical activity to the visual cortex, by means of TTX intra-vitreal injections, did not result in a modulation of TrkA expression in this area.

This result indicates that the developmental up-regulation of TrkA observed is not due to the increase of electrical input that takes place at the time of eye opening, as postulated before. A similar result was obtained in the study of p75^{NTR} expression: the visual input does not modulate p75^{NTR} expression in the visual cortex. All the same, it must be born in mind that TTX intra-vitreal injections do not eliminate spontaneous intra-cortical activity or activity from other afferent systems. Understanding whether or not NGF receptor expression in the rat visual cortex is dependent upon electrical activity, needs more investigations.

NGF does not modulate NGF receptor expression in the visual cortex

Previous studies showed that NGF expression in the rat cortex increases substantially in the postnatal period, and in particular during the critical period for cortical plasticity (Large *et al.*, 1986); this increase takes place during the same period of cortical TrkA up-regulation detected in this study.

I decided to investigate whether NGF can modulate the efficacy of its own action on cortical plasticity by modulating the level of expression of its receptors TrkA and p75^{NTR}. From the results obtained in this study it can be concluded that application of NGF to the visual cortex does not affect NGF receptors proteins expression in the rat visual cortex. In their study, Li *et al.* (1995) showed that intra-ventricular injections of NGF, that is a differentiative and survival factor for BFCN (Thoenen *et al.*, 1987; Hefti *et al.*, 1989), are able to induce an increase of TrkA mRNA levels in these neurons; according to these authors, this increase would be too small to be detected also at protein level. This conclusion could be valid also for the cortical TrkA expression detected in the visual cortex in this study.

Sites of cortical NGF receptor expression: BFCN terminals, cortical neurons and other afferent systems

Characterising where NGF receptors are expressed at cortical level during the critical period would help identify the sites of NGF action in cortical plasticity. Because of the negative results obtained in the immunohistochemical experiments, we cannot know the precise cellular localisation of cortical NGF receptors. Anyway, the effects observed following the lesion of BFCN, together with the results previously obtained in the study of NGF-induced neurotransmitter release and the findings from other groups, allow us to draw some plausible conclusions: 1) BFCN terminals present in the visual cortex bear both TrkA and p75[NTR] receptors; 2) besides BFCN terminals, other neurons/terminals in the visual cortex, express NGF receptors.

1) BFCN are projecting neurons whose axons extend throughout neo-cortical areas; in particular the visual cortex receives cholinergic input from the hdbB, that resides in the rostral part of the basal forebrain, and from the nbM, which is more caudal (Carey and Rieck, 1987; Dinopoulos *et al.*, 1989). The survival and differentiation of these neurons are dependent upon NGF that is retrogradely accumulated from the target area of innervation and that is also able to potentiate cholinergic transmission (Thoenen *et al.*, 1987; Hefti *et al.*, 1989; Rylett and Williams, 1994). Some groups showed that administration of radioactive NGF to the visual cortex results in retrograde labelling of cholinergic basal forebrain nuclei and not of other nuclei (Seiler and Schwab, 1984; Domenici *et al.*, 1994). Moreover, Pizzorusso *et al.* (1999) showed that TrkA agonist, RTA, administered to the visual cortex, is transported to these nuclei. Finally, cortical p75[NTR]-immunoreactivity coincides with the laminar distribution of cholinergic afferents (Pioro and

Cuello, 1990). All these experiments suggest that axons of BFCN terminating in the visual cortex express NGF receptors.

A possible strategy in order to investigate directly whether cortical cholinergic terminals bear NGF receptors was, then, to selectively eliminate these fibres. After the localisation of the nuclei that selectively project to the visual cortex, I treated P16 animals with quisqualic acid injections in order to eliminate these nuclei. Because the aim of the present study was to damage as much as possible the cholinergic afferents to the visual cortex I used large doses of this drug (see Materials and Methods). All the effects of the lesion were mainly present in the ipsilateral injected side, but, although to a lesser extent, significant effects were detected also on the contralateral side, possibly due to diffusion of the high volume of the drug solution injected.

The lesion of BFCN has been largely used as a useful experimental approach to investigate the role of cholinergic system in the brain (Mesulam, 1995). The main drugs used for such a lesion are excitotoxic drugs (Dunnett et al., 1991), such as the one used in this study (quisqualic acid), and other neurotoxins (such as 192-IgG saporin; Wiley et al., 1991). Quisqualic acid has the largest excitotoxic effect among the different drugs/toxins used for such lesions. This drug acts through non-NMDA glutamate receptor, then it damages only cellular structures present in the injected area that express these receptors; cellular bodies, where these receptors are richly expressed, are damaged, whereas fibres passing through the injected area, where non-NMDA receptors are practically absent, are spared by the lesion.

The efficacy of cholinergic system lesions is usually verified by analysing the presence and morphological features of BFCN in the injected area (histologically or by immunolabelling proteins expressed in these neurons, such as ChAT or NGF receptors). At cortical level, the expression or the activity of different enzymes involved in acetylcholine turnover is measured as reliable markers for cholinergic terminals presence. The main are: AchE (the enzyme responsible for acetylcholine degradation) and ChAT activity (acetylcholine production). An additional technique is the measure of choline uptake (that represents the rate limiting step of acetylcholine synthesis).

Many studies report that such lesions induce an almost complete elimination of BFCN in the injected area and also of AchE-positive fibres at cortical level. The maximum decrease of ChAT-activity observed after the lesion of the cholinergic system in the rat cortex is about 70%. It has been suggested that the activity residual after the lesion is due to the presence of cortical cholinergic interneurons (Johnston *et al.*, 1981; Houser *et al.*, 1985). It must be admitted that the existence of such interneurons in the rat cortex is still debated: ChAT-immunoreactivity has been detected in the rat cortex, but many groups failed to detect its mRNA. However, by using an extremely sensitive technique (single-cell RT-PCR), Cauli *et al.* (1997) recently showed that some interneurons expressing ChAT mRNA exist at cortical level.

The maximal decrease of choline uptake observed at cortical level after the lesion is about 40% (Arendash *et al.*, 1987; Roßner *et al.*, 1995). A total decrease in this parameter is not to be expected after the lesion. Indeed, as already mentioned, the lesion does not affect cortical cholinergic interneurons; moreover, it has been observed that choline uptake is not only a feature of cholinergic terminals, but, also of other neurotransmitter systems (Pittaluga and Raiteri, 1987). All these structures can account for the residual choline uptake observed after the lesion.

In this study I observed that quisqualic acid injections resulted in a strong reduction of basal forebrain ChAT-immunoreactivity and cortical AchE. Measurements of ChAT-activity and choline uptake at cortical level were found to decrease to an extent comparable to the one obtained by other groups. These results indicate a selective and strong efficacy of the lesion protocol used. Pilot experiments indicated that BFCN lesions obtained with the neurotoxin 192-IgG saporin did not decrease ChAT-activity more severely than quisqualic acid, indicating a similar efficacy for the two drugs.

Finally, I found that in the visual cortex of quisqualic acid treated animals the expression of both TrkA and p75^{NTR} receptors strongly decreases. From these results I concluded that both TrkA and p75^{NTR} receptors are present on the terminals eliminated by the lesion, that is, cholinergic terminals originating from BFCN.

The presence of NGF receptors on cortical cholinergic terminals from BFCN can account in part for the developmental up-regulation of TrkA detected at cortical level in this study. These fibres are already present in the cortex around birth, and during the first weeks of postnatal life their axonal plexus gradually increases and, approximately at the fourth week of life, reaches an adult-like level. Li *et al.* (1995) have recently shown that TrkA expression increases in the cellular bodies of BFCN from P4 to P21. The development of BFCN and of their innervation of the cortex is concomitant to the period of cortical TrkA receptor up-regulation detected in this study. It is possible, then, that during the formation of these terminals, BFCN increase the synthesis and/or the translocation of TrkA receptor from the cellular bodies to their synaptic terminals, as already suggested for the hippocampal area that is innervated by septal cholinergic neurons (Li *et al.*, 1995).

Concerning p75[NTR], I found that its expression is not modulated during cortical development. It is generally thought that, unlike interneurons, the majority of projecting neurons, such as BFCN, coexpress TrkA and p75[NTR] (indeed p75[NTR] seems to facilitate the retrograde signalling of the ligand/receptor complex).

As a consequence, it is possible that in cortical cholinergic terminals the expression of TrkA and p75[NTR] is differentially regulated. Moreover, the possibility that the two receptors do not completely co-localise at cortical level cannot be excluded.

2) TrkA and p75[NTR] protein expression is dramatically reduced in the visual cortex of BFCN lesioned animals. However, a residual cortical NGF receptors signal, although minimal, was still detected

after the lesion. Even if the origin of this residual signal is unknown, some possible consideration can be drawn.

As BFCN projecting to the visual cortex are spread along the basal forebrain area and not packed in a single area, it is possible that the lesion spares some of their cortical terminals; the presence of these residual terminals could account for the NGF receptors residual signal detected in the visual cortex after the BFCN lesion. All the same, this conclusion, although possible, seems to be unlikely. Indeed, the experimental procedure for the BFCN lesion used in this study is considered one of the most severe; moreover, all cholinergic markers analysed for detecting the efficacy of this lesion strongly decrease.

Another possibility is that the residual signal originates from cortical neurons and/or other cortical terminals (different from BFCN terminals).

The presence of NGF receptors, and in particular TrkA, at the level of the visual cortex has been long debated. However, results obtained by using different experimental approaches suggest that some cortical neurons express this receptor. Indeed, TrkA mRNA expression has been detected in the visual cortex by some laboratories, although at a very low level (Miranda *et al.*, 1993; Valenzuela *et al.*, 1993; Cellerino *et al.*, 1996). It must be admitted that other laboratories failed in detecting TrkA mRNA expression in the visual cortex (Merlio *et al.*, 1992; Gibbs and Pfaff, 1994; Schoups *et al.*, 1995). Unfortunately, there are no data on p75^{NTR} mRNA expression at the level of the visual cortex. Another evidence comes from the group of L. Katz, who recently showed that NGF administration to ferret visual cortical slices, at a stage in which the cortex is still immature, induces a morphological modification of the dendritic arborisation in specific pyramidal neurons (McAllister *et al.*, 1995), thus indicating the presence of NGF receptors on these neurons. It has also been observed that a little amount of radioactive NGF is transported intracortically (Domenici *et al.*, 1994).

Further evidence comes from our previous results on NGF potentiating effects on presynaptic neurotransmitter release (Sala *et al.*, 1998). We found that NGF is able to potentiate the release of acetylcholine and glutamate from visual cortical synaptosomes during the critical period for plasticity, with TrkA playing the major role and $p75^{NTR}$ a minor positive role. NGF-responsive cholinergic synaptosomes could originate from BFCN cortical terminals and/or from cortical cholinergic interneurons bearing NGF receptors. The possibility that the NGF-induced glutamate release occurred indirectly through acetylcholine results unlikely, as the effect of NGF on glutamate release was insensitive to muscarinic and nicotinic acute receptor blockade. Moreover, the characteristics of the technique employed, that is, a thin layer of synaptosomes up-down superfused under conditions with no biophase, in which indirect effects are minimised, allowed to conclude that NGF added to the superfusion medium acted directly on NGF receptors-bearing glutamatergic terminals present in the visual cortex (Raiteri *et al.*, 1974; Raiteri *et al.* 1984). The possibility that these terminals may originate from the main extracortical glutamatergic input to the visual cortex, that is from dLGN thalamic afferents, is unlikely; indeed many groups failed to detect the expression of both mRNA and protein for NGF receptors in the dLGN. Moreover, different studies indicated that neither radioactive NGF nor TrkA agonist (RTA) is transported from the cortex to the dLGN. It is, then, likely that glutamatergic terminals responsive to NGF originates from cortical neurons.

All these experimental evidences suggest that the residual NGF receptors signal detected in my experiments following BFCN lesion, could arise from cortical cholinergic interneurons and from glutamatergic neurons, which indeed are spared by the lesion.

Concerning other afferent systems, it has been observed that NGF receptors are expressed in the raphe serotonergic nucleus (Sobreviela *et al.*, 1994), whose cortical projections constitute another potent modulator of cortical plasticity. Serotonergic terminals are present also in the visual cortex, and could account, although to a small extent, for the presence of NGF receptors at cortical level. As in this study NGF effects on serotonine release were not measured, the possibility that the serotonergic system is a target of NGF action cannot be excluded.

Cholinergic system lesion down regulates cortical NGF responsiveness

I decided to verify the functionality of NGF receptors residual after BFCN lesion by measuring NGF effects on neurotransmitter release form visual cortical synaptosomes. For coherence with the results obtained in total protein extract samples, in this study I investigated NGF receptor expression also in synaptosomes prepared from P30 rat visual cortices. I verified that NGF receptors, TrkA and p75^{NTR}, are both present in the synaptosomal preparation and that, in lesioned animals, their expression decreases to an amount similar to the one observed in total protein extracts. Finally, I found that in lesioned animals NGF does not induce the K$^+$-evoked release of glutamate and acetylcholine from visual cortical synaptosomes anymore.

In our previous work we observed that the NGF-induced glutamate release from cortical synaptosomes of normal animals is not affected by pharmacologically blocking acetylcholine receptors (Sala *et al.* 1998). This result is not in contrast with the result obtained in the present work. Indeed, although the pharmacological blockade of cholinergic receptors represents a selective and specific experimental approach, it must be remembered that it was performed acutely on an *in vitro* preparation. On the contrary, cholinergic system lesion was performed on living animals that were sacrificed two weeks after the injections, the lapse of time needed for the complete transneuronal degeneration of BFCN projecting fibres. It must be noted that during this period, elimination of one of the most important neuromodulatory input for cortical development can induce the activation of many rearrangements and compensatory mechanisms in target areas. For instance, it has been shown that similar lesions induced an alteration in the expression and functionality of many neurotransmitter receptors at cortical level (Roßner *et al.*, 1995).

Our interpretation of the result obtained is that BFCN lesion blocked NGF potentiating effect on glutamate release not because NGF-induced acetylcholine is the mediator in this phenomenon (indeed NGF normally acts directly on glutamatergic terminals), but because the absence of a normal cholinergic input during cortical development determines a strong negative modulation of NGF responsiveness in glutamatergic terminals. The normal level of glutamate content and the normal response to depolarisation observed in the synaptosomal preparation from lesioned animals, as compared to that from PBS-injected and untreated animals, indicate that the lesion does not eliminate these terminals, but that they are vital and well functioning. As a consequence, their inability to respond to NGF could more probably reflect a lesion-induced down-regulation of NGF receptor expression in these terminals and/or a damage in one or more of the proteins involved in the NGF signalling.

A similar conclusion can also be drawn for the observed effect of the lesion on NGF-induced acetylcholine release. However, we cannot distinguish acetylcholine release from the two possible NGF cholinergic targets, BFCN cortical terminals and cholinergic interneurons. As a consequence, we do not know whether cholinergic interneurons, which are spared by the lesion, do not respond physiologically to NGF administration, or whether the lesion impairs their NGF responsiveness.

The effects of NGF on both acetylcholine and glutamate release from the rat visual cortex were completely blocked following BFCN lesion. Theoretically such an effect could be

54

explained by the elimination of a co-transmitter system present in the visual cortex that uses concomitantly acetylcholine and glutamate. Docherty *et al.* (1987) indeed postulated that glutamate could have a co-transmitter role in cortical cholinergic terminals. Contrary to this hypothesis, several studies demonstrated that following the lesion of the BFCNs projecting to the cortex, cholinergic transmission markers are strongly reduced, whereas glutamate content and release are completely unaffected (Szerb and Fine, 1989). These results clearly go against the co-transmitter role of glutamate. The presence of a very small fraction of terminals in which glutamate plays such role cannot be completely excluded; however, its presence would not account for the majority of glutamatergic and cholinergic transmission at cortical level.

Concluding remarks: possible sites and mechanisms of NGF action in cortical plasticity

In their initial study Maffei and collaborators showed that exogenous NGF administration during the critical period for plasticity prevents the effects of monocular deprivation in the rat visual cortex (Maffei *et al.*, 1992). At that time, the sites and the mechanisms of NGF action were completely unknown. Today, the large number of studies performed in Maffei's and others' laboratories indicate some plausible sites and mechanisms for NGF action on visual cortical plasticity, as simply depicted in Figure 22.

In monocularly deprived animals the decrease in synaptic efficacy of deprived fibres can be prevented by NGF through a direct action on visual neurons but also through an action on cholinergic terminals. A direct NGF effect on glutamatergic geniculo-cortical terminals seems unlikely, because LGN does not express NGF receptors and does not retrogradely transport radioactive NGF from the cortex.

NGF action on cortical plasticity would result from an increase of cortical excitatory transmission; indeed NGF was found to potentiate the release of acetylcholine and glutamate but not of GABA. An augmented cholinergic and glutamatergic input is expected to facilitate depolarising responses in visual cortical cells (Sillito and Kemp, 1983; Sato et al., 1987; Carmignoto et al., 1997). Higher responsivity of cortical units to the visual input would allow also less active, deprived eye afferents to reach the threshold required for postsynaptic activation and therefore to become more effective in driving the cortical targets.

All our experiments suggest that the cholinergic input plays a key role in mediating NGF effects at cortical level. Indeed, elimination of cholinergic input to the visual cortex completely eliminates NGF potentiating effects on glutamate and acetylcholine release. The lesion of the cholinergic system resulted not only in a strong decrease of cortical NGF receptor expression, but also in a possible modulation of the NGF signalling machinery on target neurons. As a consequence, the cholinergic system would seem to be necessary for NGF preventing effects on monocularly deprived animals.

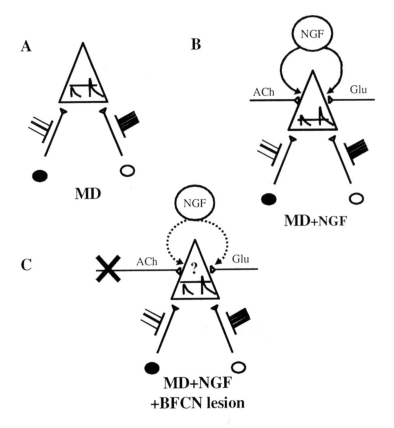

Figure 22. A schematic model for NGF actions in cortical plasticity. (A) A cortical neuron (triangle) is represented receiving input from the normal and the deprived eye. Only inputs related to the open eye are strong enough to reach the threshold for postsynaptic activation, while the deprived eye is ineffective in driving the cortical cells. (B) Infused NGF acts on both cholinergic terminals arising from the basal forebrain and intracortical glutamatergic terminals, leading to an increased acetylcholine and glutamate release. This may lower the threshold for postsynaptic activation so that weak inputs from the closed eye are now able to drive the cortical neurons. (C) Lesion of cholinergic input to the visual cortex results in elimination of NGF potentiating effects on glutamate and acetylcholine release, through a strong decrease of cortical NGF receptor expression, and possibly through modulation of the NGF signalling machinery on target neurons. According to these findings NGF would no more prevent MD effects in absence of the cholinergic input.

A further study of NGF effects on monocularly deprived animals also subjected to cholinergic system lesion, will help verify whether the above suggested site and mechanism of NGF action on visual cortical plasticity are valid also *in vivo*.

MATERIALS AND METHODS

MATERIALS AND METHODS

Animals: For western blot analysis, Long Evans hooded rats (Charles River, Italy) of different postnatal ages were anaesthetised with chloral hydrate (intraperitoneal injection, 4 ml/kg of a 0.64 M solution) and decapitated. Brain areas of interest were immediately removed and immersed at 4°C in lysis buffer for homogenisation. For immunohistochemistry, animals were anaesthetised, transcardially perfused with saline solution followed by 4% paraformaldehyde in 0.1 M phosphate buffer solution (PBS). The brains were removed, rinsed overnight (o/n) in 30% sucrose in PBS for cryoprotection and cut coronally at 40 μm with a freezing microtome. For the study of neurotransmitter release animals were killed by decapitation and tissues rapidly removed (see Synaptosomal preparation).

RTA: the production and specificity of this IgG antibody have been previously described (Clary et al., 1994). Briefly, a polyclonal antiserum was prepared that recognises the entire extracellular domain of the TrkA receptor (aminoacids 1-416). A truncated portion of the full-length rat TrkA coding sequence containing the extracellular domain was generated by polymerase chain reaction (PCR). The expression cassette directed the production of a truncated TrkA protein in baculovirus-infected Spodoptera frugiperda cells (Sf900). RTA antibody does not recognise either TrkB or TrkC.

Fab: this fragment corresponds to the antigene-binding domain of RTA IgG antibody, and it is obtained from RTA with standard procedure (Clary et al., 1994).

anti-TrkA (Biolabs): anti-TrkA antibody (New England Biolabs) is produced by immunising rabbits with a synthetic peptide (KLH coupled) corresponding to residues 777 to 790 of human TrkA. It is purified by protein A-chromatography; the resulting IgG fraction is further purified by elution from a non-phosphopeptide affinity column.

anti-p75^{NTR}: anti-human p75^{NTR} from Promega is a rabbit polyclonal antibody raised against the recombinant fusion protein (human p75^{NTR} cytoplasmic domain) produced in Escherichia Coli and purified using protein G.

anti-ChAT: the anti-ChAT antibody (mab305) from Chemicon Inc. is a mouse monoclonal antibody raised against choline acetyltransferase purified from rat brain.

NGF: it was purified from mouse submaxillary gland and kindly provided by Dr. D. Mercanti (Neurobiology CNR, Rome, Italy).

anti-β-tubuline: the anti-β-tubuline antibody from Sigma is a mouse monoclonal antibody raised against β-tubuline obtained from fluid ascites.

Immunohistochemistry

After several rinses in PBS, sections were incubated for two hours in agitation at room temperature (RT) in a blocking solution (2% normal horse serum- NHS, 0.3% Triton X-100 in PBS for anti-

ChAT; 10% normal goat serum - NGS, 0.3% Triton X-100 in PBS for RTA). Then, sections were incubated in the primary antibody solution (anti-ChAT 1:3000 in the relative blocking solution; RTA 1:1000 in 2% NGS, 0,3% Triton X-100 in PBS). After washing the tissue in PBS (3x10 min), sections were incubated for two hours at RT in the secondary antibody solution (horse anti-mouse 1:200 in 1% NHS, 0.1% Triton X-100 in PBS for anti-ChAT; goat anti-rabbit 1:200 in 5% NGS, 0.1% Triton X-100 in PBS). Both secondary antibodies are biotin conjugated (Vector). Following 3x10 min rinses in PBS, sections were incubated for 1 hour in an avidin-biotin complex (Elite kit, Vector Laboratories; 1:125). After rinsing in PBS 3x10 min and once in acetate buffer 0.1 M pH 6.0, the reaction was developed by diaminobenzidine/nichel method. Sections were mounted on gelatine-coated slides, dehydrated through graded alcohol, cleared in xylenes and coverslipped with Permount.

Histochemistry

Brain sections were washed in 0.1 M acetate buffer pH 5.1 for 5 min and then incubated in the same solution containing 2 mM glycine, 2 mM $CuSO_4$, 80 mM $MgCl_2$, 1 mg/ml acetylcholine iodide and 0.1 mM ethopropazine for 3 hrs at 38°C. After a 10 min wash in 0.1 M PB, sections were incubated in 1% ammonium sulphide in 0.1 M PB for 3 min. Following a 10 min wash in 0.9% NaCl, sections were incubated in 0.2% AuCl for 7 min. The reaction was stopped by an incubation in sodium thiosulfate for 5

60

min, and the sections were mounted on gelatine-coated slides, dried, dehydrated and coverslipped with Permount.

Western blot analysis

Proteins were extracted from brain areas of interest according to Knüsel *et al.* (1994). Briefly, samples were homogenised in 100-200 μl of lysis buffer (150 mM NaCl, 20 mM Tris pH 8.0, 10% glycerol, 1% Triton X-100, 1 mM PMSF, 1 mg/ml aprotinin, 1 mg/ml leupeptin, 1 mM sodium orthovanadate) and incubated for 30 min in ice. Lysates were centrifuged (13000 rpm for 30 min at 4°C) to eliminate cellular debris, and protein concentration in the supernatants was estimated with the Bradford method (Biorad protein assay kit). Then, samples were boiled in loading buffer, electrophoresed on 10% SDS-polyacrilamide (PAGE) minigel and transferred to nitro-cellulose (Amersham). Blots were incubated in a blocking solution (4% non-fat dry milk with 0.2% Tween-20 in Tris Buffered Solution, TBS) for 2 hrs at RT and then probed o/n at 4°C with primary antibody (1μg/ml of RTA, anti-TrkB or anti-p75[NTR] antibody in a 0.1% Tween-20 TBS solution containing 2% non-fat dry milk). After several washes, blots were incubated with HRP labelled anti rabbit secondary antibody (Biorad, 1:3000) for 2 hours at 30ºC and analysed using ECL chemiluminescence system (Amersham).

To estimate the level of neurotrophin receptor expression detected in the western blot experiments, filters were washed for 30 min at 50°C in 62.5 mM Tris pH 6.8, 2% SDS, 100 mM βmercaptoethanol, incubated in blocking solution (4% Bovine Serum Albumine - BSA, with 0.2% Tween-20 in TBS) and then re-probed with anti-β-tubuline antibody (Sigma) diluted in a 0.1% Tween-20 TBS solution containing 2% BSA. Signal was then revealed by using a HRP labelled anti mouse secondary antibody (Biorad, 1:5000) and ECL.

Immunoprecipitation

Lysates (1 mg of total protein extracts in 0.5 ml of lysis buffer) were immunoprecipitated with 2 μg of anti-TrkA (Biolabs) while rocking on a rotating wheel for 2 hours at 4°C. Immunoprecipitates were collected at 4°C by incubating with protein A-Sepharose beads (30 μl of a 1:1 solution in TBS; CL-4B Pharmacia Biotech) for 2 hours. After several washes with TBS, Sepharose-bound proteins were eluted in loading buffer and processed for SDS-PAGE immunoblot analysis.

Densitometry on western blots

To assess semi-quantitatively the different signals obtained in western blot analysis, several sheets of x-ray film were exposed to each blot for varying lengths of time between 1 and 15 min. The bands of the developed films were quantified utilising the MCID Image Analysis System. This system identifies objects within a user-defined window, measures the brightness of each pixel and the total area of the objects, and calculates their mean optical densities (O.D.). A window size was chosen to include one band for each measurement. For each band an index of the precipitated silver in the film's emulsion was calculated as the product of mean O.D. and area of the band. Averaging these values from each film exposed to the same blot, and corrected by the exposure time of the film, a relative amount of the signal was estimated for each sample. Only values within the linear range of the film were used for this calculation. The linear range of the film was calculated loading an increasing amount of both PC12 and visual cortex samples; in each blot a PC12 sample was loaded as positive control. In order to compare the signal obtained in samples from different visual cortices, the same amount of total protein extracts for each sample was loaded in each lane (i.e. 200 µg for visual cortex).

O.D. of neurotrophin receptors and β-tubuline bands were calculated as described above. Ratios of neurotrophin receptors/ β-tubuline O.D. values (mean±SEM of different experiments) were calculated and plotted on a graph.

Tetrodotoxin injections

Tetrodotoxin (TTX, Sigma), a Na^+ channel blocker, was administered by intravitreal injection with a pulled micropipette connected to a microinjector. The micropipette was inserted at the *ora serrata* and the injection volume slowly released in the vitreous. TTX (1-2 µl of a 3.5 mM solution in 0.05 M citrate buffer, pH 4.8) was injected in the right eye. As control, the left eye was injected with the same citrate buffer vehicle solution. The same injection protocol used for the TTX treated eye was adopted for the control eye. The pupillary response to illumination was used to monitor TTX effect. Two pools of animal treatments were used: in the chronic pool, animals were treated with a single TTX injection every 24-36 hours, for 7 days, from P23 to P30; in the short-term pool, P30 animals were subjected to only one injection and sacrificed 12 hrs later. In both cases, visual cortices were explanted after a maximum period lapse of 12 hours.

Osmotic minipump implantation

Nerve Growth Factor (NGF) was directly infused into the visual cortex by means of a cannula-minipump system. Rats were anaesthetised with intraperitoneal avertin (1ml/Kg) and placed in a stereotaxic frame for minipump implantation. Miniature minipumps (Alzet 1007D, Alza, USA; pumping rate 0.5 μl/hr) were filled with NGF or cytochrome C (1μg/μl in sterile saline) and connected with polyethylene tubing to 30-gauge stainless steel cannulae. A small hole was made in the skull (1 mm lateral and in correspondence with lambda) and the cannula was lowered into the cortex. The minipump was positioned subcutaneously under the neck and the cannula secured to the skull with acrylic cement. After the dental acrylic had hardened, the scalp was sutured.

Stereotaxic coordinates for BFCN lesion

The correct stereotaxic coordinates for selectively lesioning the two nuclei that project specifically to the visual cortex were previously established by Siciliano *et al.* (1997). Briefly, a specific neuronal tracer Wheat Germ Agglutinin-Horse Radish Peroxidase conjugated (WGA-HRP Sigma, St. Louis, MO; 100 nl, 4% in saline solution) was injected by a glass micropipette in the basal forebrain region at different coordinates with respect to bregma in anaesthetised P16 animals. After a survival time of two days, the animals were transcardially perfused; then, sections were processed with a staining procedure. The correct coordinates are 0.5 mm posterior to bregma, 2.7 mm from the midline and 7.4 mm from the pial surface for nucleus basalis Magnocellularis (nbM); 0.8 mm anterior to bregma, 0.8 mm from the midline and 7.7 mm from the pial surface for horizontal limb of the diagonal band of Broca (hDBB). The vital dye, blue pontamine, was injected with the same procedure at those stereotaxic coordinates.

Quisqualic acid (Sigma, 0.5-1 µl) 0.16 M dissolved in 0.1 M PBS (pH 7.4) was unilaterally injected in both the nbM and the hdbB of anaesthetised rats using a glass micropipette connected to a microinjector. Each infusion lasted 3 min, and an additional 3 min were allowed for diffusion before the pipette was removed. Quisqualic acid is a non-NMDA glutamate receptor agonist and has an excitotoxic effects on cells expressing these receptors. In a couple of weeks, quisqualic acid induces also a transneuronal degeneration of projecting fibres. Pilot lesion experiments were also performed by using the neurotoxin IgG-192 saporin (Chemicon) at different doses (from 20 to 400 ng per injection site) with the same experimental procedure as described above.

Synaptosomal preparation

Animals were killed by decapitation and visual cortices rapidly removed, the tissue was weighted and processed to obtain crude synaptosomes according to Gray and Wittaker (1962) with minor modification. Briefly, the tissue was homogenised at 4°C in 40 volumes of 0.32 M sucrose (buffered at pH 7.4 with phosphate) by using a glass-Teflon tissue grinder. The homogenate was centrifuged (5 min at 1000 g) and synaptosomes were isolated from supernatant by centrifugation at 12000 g for 20 min. Protein content was determined with Biorad protein assay kit. The same synaptosomal preparation was used in part for the study of neurotransmitter release, in part for analysis of lesion efficacy (choline uptake, and ChAT activity), and in part for immunoblot analysis of NGF receptor expression.

Neurotransmitter release

The synaptosomal pellet was resuspended in a physiological medium containing 125 mM NaCl, 3 mM KCl, 1.2 mM $MgSO_4$, 1.2 mM $CaCl_2$, 1.0 mM NaH_2PO_4, 22 mM $NaHCO_3$, 10 mM glucose; the solution was aerated with a 95% O_2 and 5% CO_2 mixture (pH 7.2-7.4) and incubated for 15 min at 37°C with 0.08 µM [^{3}H]choline. Identical aliquots of the synaptosomal suspension were layered on microporous filters at the bottom of parallel superfusion chambers maintained at 37°C (Raiteri *et al.*, 1974) and superfused with physiological medium (0.5 ml/min) supplemented with 0.1% dialysed BSA. After 36 min of equilibration, two 3-min samples (basal release) before (min 36-39) and after (min 45-48) one 6-min sample (evoked release, min 39-45), containing the transmitters released by high K^+, were collected. A 90 sec period of depolarisation (15 mM KCl) was applied at time = 39 min. Nerve growth factor (NGF, 100 ng/ml) was added at time = 30 min and maintained until the end of the experiments. Endogenous glutamate was measured by high-performance liquid chromatography (HPLC) analysis with fluorimetric detection, after precolumn derivatisation with o-phtalaldehyde and calculated as pmol/mg of protein. Tritium content in those superfusate fractions was expressed as fractional rate. The depolarised evoked overflow was estimated by subtracting the aminoacid (or radioactivity) content in the basal release fractions from that in the 6 min fraction collected during and after depolarisation. NGF effects were evaluated as the ratio of the depolarisation evoked overflow in the presence of NGF versus that under control condition. In each experiment synaptosomal preparation from ipsilateral and contralateral visual cortex of quisqualic acid- and PBS-injected animals were analysed along with synaptosomes prepared from visual cortex of control untreated animals.

Choline uptake

The uptake of [^{3}H]choline was studied according to the following procedure: each aliquot of synaptosomal suspension (500 μl) containing about 3 mg of freshly dissected tissue (approximately 50 μg of protein content) was pre-incubated in a rotatory thermostated water bath for 10 min at 37°C. [^{3}H]choline was then added to a final concentration of 0.3 µM and incubation continued for 2 min further. After labelling, samples were rapidly collected on Whatman glass microfibre filters (GF/B) under vacuum conditions and then washed three times in 5 ml of standard medium. Filters were counted for radioactivity. Blank values were obtained by labelling samples at 4°C. In each experiment, synaptosomes prepared from the ipsilateral and contralateral visual cortex of quisqualic acid or PBS-injected animals were analysed along with synaptosomes prepared from visual cortex of control untreated animals.

ChAT-activity determination

ChAT-activity was determined according to the radiochemical method of Fonnum (1975). Briefly, synaptosomal preparation was homogenised in 20 volumes of 10 mM EDTA buffer (pH 7.4) and 0.2% Triton X-100 at 4°C. Two µl aliquots of the homogenate were added to 5 µl of the incubation medium and then incubated for 15 min at 37°C. The composition of the incubation medium was as follows: [1-^{14}C]-acetylcoenzyme A (specific activity 51 mCi/mmol, Amersham) diluted with unlabeled compound (Boehringer Mannehim) to give finally 16.9 mCi/mmol in 0.6 mM solution, 10 mM choline chloride, 300 mM NaCl, 41 mM sodium phosphate buffer (pH 7.4), 0.1 mM physostigmine salicylate in 100 mM EDTA, and 0.05% Triton X-100. The radioactivity was determined with a liquid scintillation spectrometer. ChAT-activity was expressed as nmol/mg of protein/h. Protein content in homogenates was determined as described above.

Statistical analysis

One-way ANOVA with Tukey's *post-hoc* test or Student's two-tailed *t*-test were performed to test the significance of differences between groups.

REFERENCES

REFERENCES

Akaneya Y., Tsumoto T., Kinoshita S. and Hatanaka H. (1997). Brain-derived neurotrophic factor enhances long-term potentiation in rat visual cortex. *J. Neurosci.* **17**: 6707-6716.

Allendoerfer K.L., Cabelli R.J., Escandòn E., Kaplan D.R., Nikolics K. and Shatz C.J. (1994). Regulation of neurotrophin receptors during the maturation of the mammalian visual system. *J. Neurosci.* **14**: 1795-1811.

Arendash G.W., Millard W.J., Dunn A.J. and Meyer E.M. (1987). Long-term neuropathological and neurochemical effects of nucleus basalis lesions in the rat. *Science* **238**: 952-956.

Barbacid M. (1994). The trk family of neurotrophin receptor. *J. Neurobiol.* **25**: 1386-1403.

Bear M.F. and Singer W. (1986). Modulation of visual cortical plasticity by acetylcholine and noradrenaline. *Nature* **320**: 172-176.

Bear M.F., Kleinschmidt A., Gu Q. and Singer W. (1990). Disruption of experience-dependent synaptic modifications in striate cortex by infusion of an NMDA receptor antagonist. *J. Neurosci.* **10**: 909-925.

Bengzon J., Kokaia Z., Ernfors P., Kokaia M., Leanza G., Nilsson O.G., Persson H. and Lindvall O. (1993). Regulation of neurotrophin and trkA, trkB and trkC tyrosine kinase receptor messenger RNA expression in kindling. *Neurosci.* **53**: 443-446.

Berardi N., Cellerino A., Domenici L., Fagiolini M., Pizzorusso T., Cattaneo A. and Maffei L. (1994). Monoclonal antibodies to nerve growth factor affect the postnatal development of the visual system. *Proc. Natl. Acad. Sci. USA* **91**: 684-688.

Berardi N., Pizzorusso T., Lodovichi C. and Maffei L. (1998). Differential effects of neurotrophins on ocular dominance plasticity of the rat visual cortex. *Soc. Neurosci. Abs.* **24**: 511.11.

Berzaghi M.P., Cooper J., Castrén E., Zafra F., Sofroniew M., Thoenen H. and Lindholm D. (1993). Cholinergic regulation of brain-derived neurotrophic factor (BDNF) and nerve growth factor (NGF) but not neurotrophin-3 (NT-3) mRNA levels in the developing rat hippocampus. *J. Neurosci.* **13**: 3818-3826.

Blöchl A. and Thoenen H. (1995). Characterization of nerve growth factor (NGF) release from hippocampal neurons: evidence for a constitutive and an unconventional sodium-dependent regulated pathway. *Eur. J. Neurosci.* **7**: 1220-1228.

Bothwell M. (1995). Functional interactions of neurotrophins and neurotrophin receptors. *Annu. Rev. Neurosci.* **18**: 221-253.

Bozzi Y., Pizzorusso T., Cremisi F., Rossi F.M., Barsacchi G. and Maffei L. (1995). Monocular deprivation decreases the expression of messenger RNA for brain-derived neurotrophic factor in the rat visual cortex. *Neurosci.* **69**: 1133-1144.

Cabelli J., Hohn A. and Shatz C.J. (1995). Inhibition of ocular dominance column formation by infusion of NT-4/5 or BDNF. *Science* **267**: 1662-1666.

Cabelli J., Tavazoie S. and Shatz C.J. (1995). Changing patterns of BDNF and NT-4/5 immunoreactivity during visual system development. *Soc. Neurosci. Abstr.* **21**: 706.16.

Cabelli J., Shelton D.L., Segal R.A. and Shatz C.J. (1997). Blockade of endogenous ligands of TrkB inhibits formation of ocular dominance columns. *Neuron* **19**: 63-76.

Caleo M., Lodovichi C. and Maffei L. (1999). Effects of TrkA activation in visual cortical plasticity require afferent spontaneous electrical activity. *Soc. Neurosci. Abs.* **23**: 457.12.

Canossa M., Griesbeck O., Berninger B., Campana G., Kolbeck R. and Thoenen H. (1997). Neurotrophin release by neurotrophins:

implications for activity-dependent neuronal plasticity. *Proc. Natl. Acad. Sci. USA* **94**: 13279-13286.

Carey R.G. and Rieck R.W. (1987). Topographic projections to the visual cortex from the basal forebrain in the rat. *Brain Res.* **424**: 205-215.

Carmignoto G., Comelli M.C., Candeo P., Cavicchioli L., Yan Q. and Maffei L. (1991). Expression of NGF receptor and NGF receptor mRNA in the developing and adult rat retina. *Exp. Neurol.* **111**: 302-311.

Carmignoto G., Canella R., Candeo P., Comelli M.C. and Maffei L. (1993). Effects of NGF on neuronal plasticity of the kitten visual cortex. *J. Physiol. (London)* **464**: 343-360.

Carmignoto G., Pizzorusso T., Tia S. and Vicini S. (1997). Brain-derived neurotrophic factor and nerve growth factor potentiate excitatory synaptic transmission in the rat visual cortex. *J. Physiol. (London)* **498**: 153-164.

Carter B.D. and Lewin G.R. (1997). Neurotrophins live or let die: does p75^{NTR} decide? *Neuron* **18**: 187-190.

Casaccia-Bonnefil P., Carter B.D., Dobrowsky R.T. and Chao M.V. (1996). Death of oligodendrocytes mediated by the interaction of nerve growth factor with its receptor p75. *Nature* **383**: 716-719.

Cases O., Vitalis T., Seif I., DeMaeyer E., Sotelo C. and Gaspar P. (1996). Lack of barrels in the somatosensory cortex of monoamine oxidase A-deficient mice: role of a serotonin excess during the critical period. *Neuron* **16**: 297-307.

Castrén E., Zafra F., Thoenen H. and Lindholm D. (1992). Light regulates expression of brain-derived neurotrophic factor mRNA in the rat visual cortex. *Proc. Natl. Acad. Sci. USA* **89**: 9444-9448.

Cauli B., Audinat E., Lambolez B., Angulo M.C., Ropert N., Tzuzuki K., Hestrin S. and Rossier J. (1997). Molecular and physiological diversity of cortical non pyramidal cells. *J. Neurosci.* **17**: 3894-3908.

Celada P., Siuciak J.A., Tran T.M., Altar C.A. and Tepper J.M. (1996). Local infusion of brain-derived neurotrophic factor modifies the firing pattern of dorsal raphe serotonergic neurons. *Brain Res.* **712**: 293-298.

Cellerino A. and Maffei L. (1996). The action of neurotrophins in the development and plasticity of the visual cortex. *Prog. Neurobiol.* **49**: 53-71.

Cellerino A., Maffei L. and Domenici L. (1996). The distribution of brain-derived neurotrophic factor and its receptor trkB in parvalbumin-containing neurons of the rat visual cortex. *Eur. J. Neurosci.* **8**: 1190-1197.

Chao M.V. (1994). The p75 neurotrophin receptor. *J. Neurobiol.* **25**: 1373-1385.

Chao M.V. and Hempstead B.L. (1995). p75 and Trk: a two-receptor system. *Trends Neurosci.* **18**: 321-326.

Choen S. (1960). Purification of a nerve-promoting protein from the mouse salivary gland and its neurocytotoxic antiserum. *Proc. Natl. Acad. Sci. USA* **46**: 302-311.

Clary D.O., Weskamp G., Austin L.R. and Reichardt L.F. (1994). TrkA cross-linking mimics neuronal responses to Nerve Growth Factor. *Mol. Biol. Cell* **5**: 549-563.

Conner J.M., Lauterborn J.C., Yan Q., Gall C.M. and Varon S. (1997). Distribution of brain-derived neurotrophic factor (BDNF) protein and mRNA in the normal adult rat CNS: evidence for anterograde axonal transport. *J. Neurosci.* **17**: 2295-2313.

Dinopoulos A., Eadie L.A., Dori I. and Paranvelas J.G. (1989). The development of basal forebrain projections to the rat visual cortex. *Exp. Brain Res.* **76**: 563-571.

Docherty M., Bradford H.F. and Wu J.-Y. (1987). Co-release of glutamate and aspartate from cholinergic and GABAergic synaptosomes. *Nature* **330**: 64-66.

Domenici L., Berardi N., Carmignoto G., Vantini G. and Maffei L. (1991). Nerve Growth Factor prevents the amblyopic effects of monocular deprivation. *Proc. Natl. Acad. Sci. USA* **88**: 8811-8815.

Domenici L., Cellerino A., Berardi N., Cattaneo A. and Maffei L. (1994). Antibodies to nerve growth factor (NGF) prolong the sensitive period to monocular deprivation in the rat. *NeuroRep.* **5**: 2041-2044.

Domenici L., Fontanesi G., Cattaneo A., Bagnoli P. and Maffei L. (1994). Nerve Growth Factor (NGF) uptake and transport following injection in the developing rat visual cortex. *Vis. Neurosci.* **11**: 1093-1102.

Dunnett S.B., Everitt B.J. and Robbins T.W. (1991). The basal forebrain cortical cholinergic system: interpreting the functional consequences of excitotoxic lesions. *Trends Neurosci.* **14**: 494-501.

Fiorentini A., Berardi N. and Maffei L. (1995). Nerve Growth Factor preserves behavioural visual acuity in monocularly deprived kittens. *Vis. Neurosci.* **12**: 51-55.

Fonnum F.A. (1975). A rapid radiochemical method for the determination of choline acetyltransferase. *J. Neurochem.* **24**: 407-409.

Friedman W.J., Black I.B. and Kaplan D.R. (1998). Distribution of the neurotrophins Brain-derived neurotrophic factor, Neurotrophin-3, and Neurotrophin-4/5 in the postnatal rat brain: an immunocytochemical study. *Neurosci.* **84**: 101-114.

Gibbs R.B. and Pfaff D.W. (1994). In situ hybridisation detection of trkA mRNA in brain: distribution, colocalization with p75[NGFR] and up-regulation by Nerve Growth Factor. *J. Comp. Neurol.* **341**: 324-339.

Gray E.G. and Whittaker V.P. (1962). The isolation of nerve endings from brain: an electron microscopic study of cell fragments derived by homogenisation and centrifugation. *J. Anat.* **96**: 79-90.

Griesbeck O., Blöchl A., Carnahan J.F., Nawa H. and Thoenen H. (1995). Characterization of Brain-derived neurotrophic factor (BDNF) secretion from hippocampal neurons. *Soc. Neurosci. Abstr.* **21**: 417.12.

Gu Q. and Singer W. (1993). Effects of intracortical infusion of anticholinergic drugs on neuronal plasticity in kitten striate cortex. *Eur. J. Neurosci.* **5**: 475-485.

Hefti F., Hartikka J. and Knusel B. (1989). Function of neurotrophic factors in the adult and aging brain and their possible use in the treatment of neurodegenerative diseases. *Neurobiol. Aging* **10**: 515-533.

Hempstead B.L., Rabin S.J., Kaplan L., Reid S., Parada L.F. and Kaplan D.R. (1992). Overexpression of the trk tyrosine kinase rapidly accelerates Nerve Growth Factor-induced differentiation. *Neuron* **9**: 883-896.

Holtzmann D.M., Kilbridge J., Li Y., Cunningham E.T. Jr., Lenn N.J., Clary D.O., Reichardt L.F. and Mobley W.C. (1995). TrkA expression in the CNS: evidence for the existence of several novel NGF-responsive CNS neurons. *J. Neurosci.* **15**: 1567-1576.

Houser C.R., Crawford G.D., Salvaterra P.M. and Vaughn J.E. (1985). Immunocytochemical localization of choline acetyltransferase in rat cerebral cortex: a study of cholinergic neurons and synapses. *J. Comp. Neurol.* **234**: 17-34.

Huang Z.J., Kirkwood A., Porciatti V., Pizzorusso T., Bear M., Maffei L. and Tonegawa S. (1998). A precocious development of the visual cortex and visual acuity in transgenic mice overexpressing BDNF in the postnatal forebrain. *Soc. Neurosci. Abs.* **22**: 682.10.

Hubel D.H. and Wiesel T.N. (1998). Early exploration of the visual cortex. *Neuron* **20**: 401-412.

Johnston M.V., McKinney M. and Coyle J.T. (1981). Neocortical cholinergic innervation: a description of extrinsic and intrinsic components in the rat. *Exp. Brain Res.* **43**: 159-172.

Katz L.C. and Shatz C.J. (1996). Synaptic activity and the construction of cortical circuits. *Science* **274**: 1133-1138

Kaplan D.R., Martin-Zanca D. and Parada L.F. (1991). Tyrosine phosphorylation and tyrosine kinase activity of the trk proto-oncogene product induced by NGF. *Nature* **350**: 158-160.

Kasamatsu T. and Pettigrew J.D. (1979). Preservation of binocularity after monocular deprivation in the striate cortex of kittens treated with 6-hydroxydopamine. J. *Comp. Neurol.* **185**: 139-161.

Klein R., Jing S., Nanduri V., O'Rourke E. and Barbacid M. (1991). The trk proto-oncogene encodes a receptor for Nerve Growth Factor. *Cell* **65**: 189-197.

Knüsel B., Rabin S.J., Hefti F. and Kaplan D.R. (1994). Regulated neurotrophin responsiveness during neuronal migration and early differentiation. *J. Neurosci.* **14**: 1542-1554.

Large T.H., Bodary S.C., Clegg D.O., Weskamp G., Otten U. and Reichardt L.F. (1986). Nerve Growth Factor gene expression in the developing brain. *Science* **234**: 352-355.

Lein E., Hohn A. and Shatz C.J. (1995). Reciprocal laminar localization and developmental regulation of BDNF and NT-3 mRNA during visual cortex development. *Soc. Neurosci. Abs.* **21**: 706.17.

Levi-Montalcini R. (1987). The nerve growth factor 35 years later. *Science* **237**: 1154-1162.

Lewin G.R. and Barde Y.-A. (1996). Physiology of the neurotrophins. *Annu. Rev. Neurosci.* **19**: 289-317.

Li Y., Holtzman D.M., Kromer L.F., Kaplan D.R., Chua-Couzens J., Clary D.O., Knusel B. and Mobley W.C. (1995). Regulation of TrkA and ChAT expression in developing rat basal forebrain:

evidence that both exogenous and endogenous NGF regulate differentiation of cholinergic neurons. *J. Neurosci.* **15**: 2888-2905.

Lohof A.M., Ip N.Y. and Poo M.-M. (1993). Potentiation of developing neuromuscular synapses by the neurotrophins NT-3 and BDNF. *Nature* **363**: 350-353.

Maffei L., Berardi N., Domenici L., Parisi V. and Pizzorusso T. (1992). Nerve Growth Factor (NGF) prevents the shift in ocular dominance distribution of visual cortical neurons in monocularly deprived rats. *J. Neurosci.* **12**: 4651-4662.

Mamounas L.A., Blue M.E., Siuciak J.A. and Altar C.A. (1995). Brain-derived neurotrophic factor promotes the survival and sprouting of serotonergic axons in rat brain. *J. Neurosci.* **15**: 7929-7939.

McAllister A.K., Katz L.C. and Lo D.C. (1995). Neurotrophins regulate dendritic growth in developing visual cortex. *Neuron* **15**: 791-803.

McAllister A.K., Katz L.C. and Lo D.C. (1996). Neurotrophin regulation of cortical dendritic growth requires activity. *Neuron* **17**: 1057-1064.

McAllister A.K., Katz L.C. and Lo D.C. (1997). Opposing roles for endogenous BDNF and NT-3 in regulating cortical dendritic growth. *Neuron* **18**: 767-778.

Meakin S.O. and Shooter E.M. (1991). Molecular investigations on the high-affinity Nerve Growth Factor receptor. *Neuron* **6**: 153-163.

Merlio J.-P., Ernfors P., Jaber M. and Persson H. (1992). Molecular cloning of rat trkC and distribution of cells expressing messenger RNAs for members of the trk family in the rat central nervous system. *Neurosci.* **51**: 513-532.

Mesulam M.-M. (1995). The cholinergic contribution to neuromodulation in the cerebral cortex. *Seminars in the Neurosciences* **7**: 297-307.

Minichiello L., Casagrande F., Tatche R.S., Stucky C.L., Postigo A., Lewin G.R., Davies A.M. and Klein R. (1998). Point mutation in TrkB causes loss of NT-4 dependent neurons without major effects on diverse BDNF responses. *Neuron* **21**: 335-345.

Miranda R.C., Sohrabji F. and Toran-Allerand C.D. (1993). Neuronal colocalization of mRNAs for neurotrophins and their receptors in the developing central nervous system suggests a potential for autocrine interactions. *Proc. Natl. Acad. Sci. USA* **90**: 6439-6443.

Mufson E.J., Lavine N., Jaffar S., Kordower J.H., Quirion R. and Saragovi H.U. (1997). Reduction in p140-TrkA receptor protein within the Nucleus Basalis and cortex in Alzheimer's disease. *Exp. Neurol.* **146**: 91-103.

Nawa H., Pelleymounter M.A. and Carnahan J. (1994). Intraventricular administration of BDNF increases neuropeptide expression in newborn rat brain. *J. Neurosci.* **14**: 3751-3765.

Nishio T., Furukawa S., Akiguchi I., Ohnishi K., Tomimoto H., Nakamura S. and Kimura J. (1994). Cellular localization of nerve growth factor immunoreactivity in adult rat brain: quantitative and immunohistochemical study. *Neurosci.* **60**: 67-84.

Pioro E.P. and Cuello A.C. (1990). Distribution of Nerve Growth Factor receptor like immunoreactivity in the adult rat central nervous system. Effect of colchicine and correlation with the cholinergic system. –I. Forebrain. *Neurosci.* **34**: 57-87.

Pittaluga A. and Raiteri M. (1987). Choline increases GABA release in the rat hippocampus by a mechanism sensitive to hemicolinium-3. *Naunyn-Schmiederber's Arch. Pharmacol.* **336**: 327-331.

Pizzorusso T., Berardi N., Rossi F.M., Viegi A., Venstrom K., Reichardt L.F. and Maffei L. (1999). TrkA activation in the rat visual cortex by antirat trkA IgG prevents the effect of monocular deprivation. *Eur. J. Neurosci.* **11**: 204-212.

Prakash N., Cohen-Cory S. and Frostig R.D. (1996). Rapid and opposite effects of BDNF and NGF on the functional organization of the adult cortex *in vivo*. *Nature* **381**: 702-706.

Raiteri M., Angelini F. and Levi G. (1974). A simple apparatus for studying the release of neurotransmitters from synaptosomes. *Eur. J. Pharmacol.* **25**: 411-416.

Raiteri M., Bonanno G., Marchi M. and Maura G. (1984). Is there a functional linkage between neurotransmitter uptake mechanisms and presynaptic receptors?. *J. Pharmacol. Exp. Therap.* **231**: 671-677.

Riddle D.R., Lo D.C. and Katz L.C. (1995). NT-4 mediated rescue of lateral geniculate neurons from effects of monocular deprivation. *Nature* **17**: 189-191.

Roßner S., Schliebs R. and Bigl V. (1995). 192 IgG-saporin-induced immunotoxic lesions of cholinergic basal forebrain system differentially affect glutamatergic and GABAergic markers in cortical rat brain regions. *Brain Res.* **696**: 165-176.

Roerig B. and Katz L.C. (1997). Modulation of intrinsic circuits by serotonin 5-HT3 receptors in developing ferret visual cortex. *J. Neurosci.* **17**: 8324-8338.

Rossi F.M., Bozzi Y., Pizzorusso T. and Maffei L. (1999). Monocular deprivation decreases brain-derived neurotrophic factor immunoreactivity in the rat visual cortex. *Neurosci.* **90**: 363-368.

Rylett R.J. and Williams L.R. (1994). Role in neurotrophins in cholinergic-neurone function in the adult and aged CNS. *Trends Neurosci.* **17**: 486-490.

Sala R., Viegi A., Rossi F.M., Pizzorusso T., Bonanno G., Raiteri M. and Maffei L. (1998). NGF and BDNF increase transmitter release in the rat visual cortex. *Eur. J. Neurosci.* **10**: 2185-2191.

Sato H., Hata Y., Masui H. and Tsumoto T. (1987). A functional role of cholinergic innervation to neurons in the cat visual cortex. *J. Neurophysiol.* **58**: 765-780.

Schoups A.A., Elliott R.C., Friedman W.J. and Black I.B. (1995). NGF and BDNF are differentially modulated by visual experience in the developing geniculocortical pathway. *Devl. Brain. Res.* **86**: 326-334.

Seiler M. and Schwab M.E. (1984). Specific retrograde transport of Nerve Growth Factor (NGF) from neocortex to Nucleus Basalis in the rat. *Brain Res.* **300**: 33-39.

Siciliano R., Fontanesi G., Casamenti F., Berardi N., Bagnoli P and Domenici L. (1997). Postnatal development of functional properties of visual cortical cells in rats with excitotoxic lesions of basal forebrain cholinergic neurons. *Vis. Neurosci.* **14**: 111-123.

Sillito A.M. and Kemp J.A. (1983). Cholinergic modulation of the functional organization of the cat visual cortex. *Brain Res.* **289**: 143-155.

Sobreviela T., Clary D.O., Reichardt L.F., Brandabur M.M., Kordower J.H. and Mufson E.J. (1994). TrkA-immunoreactive profiles in the central nervous system: colocalization with neurons containing p75 Nerve Growth Factor receptor, Choline Acetyltransferase, and Serotonin. *J. Comp. Neurol.* **350**: 587-611.

Szerb J.C. and Fine A. (1989). Is glutamate a co-transmitter in cortical cholinergic terminals? Effects of nucleus basalis lesion and of presynaptic muscarinic agents. *Brain Res.* **515**: 214-218.

Thoenen H., Bandtlow C. and Heumann R. (1987). The physiological function of nerve growth factor in the central nervous system: comparison with the periphery. *Rev. Physiol. Biochem. Pharmacol.* **109**: 145-178.

Valenzuela D.M., Maisonpierre P.C., Glass D.J., Rojas E., Nuñez L., Kong Y., Gies D.R., Stitt T.N., Ip N.Y. and Yancopoulos G.D. (1993). Alternative forms of rat TrkC with different functional capabilities. *Neuron* **10**: 963-974.

Wang Y., Gu Q. and Cynader M.S. (1997). Blockade of serotonin-2C receptors by mesulergine reduces ocular dominance plasticity in kitten visual cortex. *Exp. Brain Res.* **114**: 321-328.

Wiley R.G., Oeltmann T.N. and Lappi D.A. (1991). Immunolesioning: selective destruction of neurons using immunotoxin to rat NGF receptor. *Brain Res.* **562**: 149-153.

Wolf D.E., McKinnon C.A., Daou M.-C., Stephens R.M., Kaplan D.R. and Ross A.H. (1995). Interaction with TrkA immobilizes gp75 in the high affinity Nerve Growth Factor receptor complex. *J. Biol. Chem.* **270**: 2133-2138.

Yan Q. and Johnson E.M. Jr. (1988). An immunohistochemical study of nerve growth factor receptor in developing rats. *J. Neurosci.* **8**: 3481-3498.

Yan Q., Radeke M.J., Matheson C.R., Talvenheimo J., Welcher A.A. and Feinstein S.C. (1997). Immunocytochemical localization of TrkB in the central nervous system of the adult rat. *J. Comp. Neurol.* **378**: 135-157.

Yeo T.T., Chua-Couzens J., Butcher L.L., Bredesen D.E. Cooper J.D., Valletta J.S., Mobley W.C. and Longo F.M. (1997). Absence of p75^{NTR} causes increased basal forebrain cholinergic neurons size, Choline Acetyltransferase activity, and target innervation. *J. Neurosci.* **17**: 7594-7605.

Zafra F., Hengerer B., Leibrock J., Thoenen H. and Lindholm D. (1990). Activity dependent regulation of BDNF and NGF mRNAs in the rat hippocampus is mediated via non-NMDA glutamate receptors. *EMBO J.* **9**: 3545-3550.

Zafra F., Castrén E., Thoenen H. and Lindholm D. (1991). Interplay between glutamate and γ-aminobutyric acid transmitter systems in the physiological regulation of brain-derived neurotrophic factor and nerve growth factor synthesis in hippocampal neurons. *Proc. Natl. Acad. Sci. USA* **88**: 10037-10041.

Elenco delle Tesi di perfezionamento della Classe di Scienze
pubblicate dall'Anno Accademico 1992/93

Donato Nicolò, *Search for neutrino oscillations in a long baseline experiment at the Chooz nuclear reactors*, 1999.

Rocco Chirivì, *LS algebras and Schubert varieties*, 2000.

Francesco Mattia Rossi, *A Study on Nerve Growth Factor (NGF) Receptor Expression in the Rat Visual Cortex: Possible Sites and Mechanisms of NGF Action in Cortical Plasticity*, 2000.

Valentino Magnani, *Elements of Geometric Measure Theory on Sub-Riemannian Groups*, 2002.

"CompoMat" Loc. Braccone, 02040 Configni (RI), Italy
Finito di stampare per conto della "CompoMat" dalla Nuova Grafica 86 nell'aprile 2004